Extreme Events of Air-Sea Interaction

Extreme Events of Air-Sea Interaction:

Case Studies from the Mediterranean Sea and the Red Sea

By

Tarek M. El-Geziry

Cambridge Scholars Publishing

Extreme Events of Air-Sea Interaction:
Case Studies from the Mediterranean Sea and the Red Sea

By Tarek M. El-Geziry

This book first published 2023

Cambridge Scholars Publishing

Lady Stephenson Library, Newcastle upon Tyne, NE6 2PA, UK

British Library Cataloguing in Publication Data
A catalogue record for this book is available from the British Library

ISBN (10): 1-5275-0124-8
ISBN (13): 978-1-5275-0124-9

CONTENTS

LIST OF FIGURES

LIST OF TABLES

LIST OF ACRONYMS

Adriatic Sea and Coast System	ASCS
African Monsoon	AM
Alexandria Eastern Harbour	AEH
Alexandria Western Harbour	AWH
Australian Monsoon	AM
Balearic RIssaga Forecasting System	BRFS
Chlorofluorocarbons	CFCs
Copernicus Marine Environment Monitoring Service	CMEMS
Coupled Model Intercomparison Project 5	CMIP-5
East	E
East Asian Monsoon	EAM
El Nino Southern Oscillation	ENSO
Empirical Orthogonal Function	EOF
European Centre for Medium-Range Weather Forecast	ECMWF
Global Tide and Surge Reanalysis	GTSR
Greenhouse Gas	GHG
Indian Monsoon	IM
Indian Summer Monsoon	ISM
Intergovernmental Panel on Climate Change	IPCC
Inter-Tropical Convergence Zone	ITCZ
Madden Julian Oscillations	MJO
Marine Atmospheric Boundary Layer	MABL
Marine Heatwave	MHW
Marine Heatwaves	MHWs
Mean Sea Level	MSL
Mid Ocean ridge	MOR
North	N

North America Mexican Monsoon System	NAMS
North Atlantic Oscillations	NAO
North Central Red Sea	NCRS
Northern Red Sea	NRS
Northeast	NE
Northwest	NW
Radius of Outermost Closed Isobar	ROCI
Red Sea Convergence Zone	RSCZ
Red Sea Trough	RST
Representative Concentration Pathway	RCP
Root-Mean Squared error	RMSE
Sea Level Rise	SLR
Sea Surface Temperature extreme	T_{max}
Sea Surface Temperature	SST
South	S
South Asian Monsoon	SAM
South Central Red Sea	SCRS
Southeast	SE
Southern America Monsoon System	SAMS
Southern Red Sea	SRS
Southwest	SW
West	W
Western Indian Ocean	WIO
World Meteorological Organization	WMO

LIST OF SYMBOLS

Actual vapor pressure	e_a
Air temperature	T_a
Argon	Ar
Carbon dioxide	CO_2
Carbon monoxide	CO
Centimeter	cm
Cloudiness	C
Daily incoming solar radiation	Q_{sc}
Degree	°
Degree Celsius	°C
hecto Pascal	hPa
Helium	He
Hour	h
Hydrogen sulfide	H_2S
Incoming solar radiation	Q_s
Kilogram	kg
Kilometer	km
Latent heat flux	Q_E
Longwave radiation flux	Q_L
Methane	CH_4
Methyl-iodide	CH_3I
Meter	m
Micrometer	µm
Millibar	mb
Millimeter	mm
Net heat flux	Q_T
Nitrogen	N_2

Nitrous oxide	N_2O
Oxygen	O_2
Ozone	O_3
Per	/
Percent	%
Saturated vapor pressure	e_s
Seawater density	ρ
Seconds	s
Sensible heat flux	Q_H
Significant wave height	H_s
Specific heat capacity	C_p
Water vapor	H_2O
Watt	W
Wave height	H
Wind speed	W
Year	yr

Chapter One

Introduction

The atmosphere, oceans, ice masses, land surface and vegetation cover all work together to create the global climate system. Nowadays, it is accepted to say that the structure of this interactive system consists of five subsystems (Houghton et al., 2001): (1) atmosphere: the body of air enveloping the earth, (2) hydrosphere: any water body found on Earth and in air, (3) cryosphere: solid water surface areas on Earth, (4) lithosphere: any inorganic material and (5) Biosphere: all living organisms.

This system (Fig. 1.1) is forced and influenced by various external forcing mechanisms, the most important of which is the sun. Also, the direct effect of human (anthropogenic) activities on the climate system is considered an external forcing (Houghton et al., 2001). Any modification imposed or implemented in one of the above-mentioned elements will have an impact on the nature and functionality of the others. As oceanographers, our interest focuses mainly on the first three subsystems; being the core of diverse oceanographic studies and research involved in the investigation of the climate change and its predictability. To investigate the air-sea interaction processes, the current work will focus solely on the first two elements listed. In fact, the atmosphere and oceans are one system that must be considered together if either is to be properly understood (Warner et al., 2010). What happens in one affects the other, and the two are connected by complex feedback loops. Therefore, this book will start by showing the principal features of these two interactive systems.

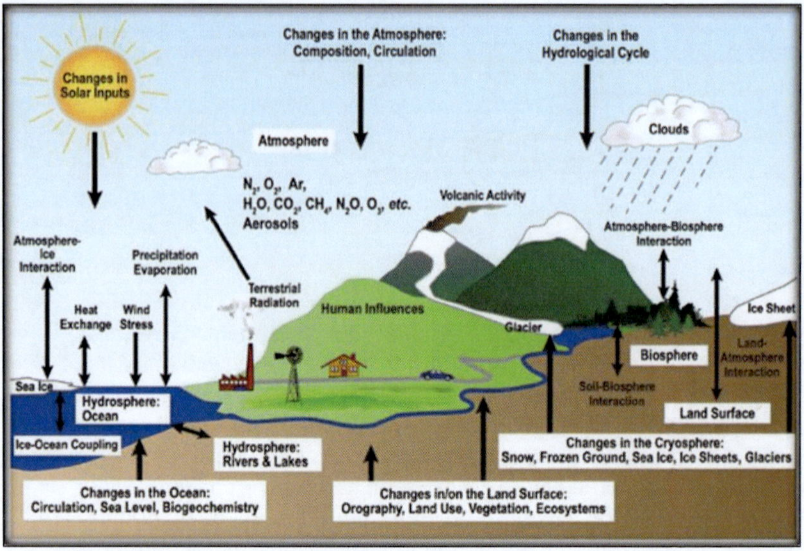

Figure 1.1. A schematic view of the global climate subsystems, their interactive processes (double-ended arrows) and some features that may change (single-ended arrows) (Source: Batisha, 2015)

1.1. Atmosphere

The earth's atmosphere is a layer of gases that surrounds the planet and is held in place by gravity (Cornish and Ives, 2009). This gaseous envelope is made up primarily of nitrogen (N_2; 78%) and oxygen (O_2; 21%), with minor amounts of other gases, including carbon dioxide (CO_2) argon (Ar), helium (He) and ozone (O_3), in addition to a variable amount of water vapor (H_2O) (Cornish and Ives, 2009). Clouds of liquid water and ice crystals are nested in the atmosphere (Ahrens, 2018). Our bodies require oxygen to burn (oxidize) food and release energy to our cells (Trujillo and Thurman, 2011). CO_2 is an important greenhouse gas (GHG) since it traps a considerable segment of the earth's outgoing energy. CO_2 is mostly released into the atmosphere through the decomposition of vegetation, but it is also released by volcanic eruptions, animal exhalations, fossil fuel combustion, and deforestation (Ahrens, 2018). Sadly to say, most of mathematical model experiments that anticipate future atmospheric conditions predict a worldwide warming of surface air between 1 °C and 3.5 °C by the year 2100 as a result of rising CO_2 (and other GHG) levels (Ahrens, 2018). The ozone layer shielded life forms from the sun's ultraviolet radiation (Wells, 2011). This layer has been affected over years, since the industrial revolution era, by the release of huge amounts of Chlorofluorocarbons (CFCs), which depleted the O_3-layer in the stratosphere leading to evident climate change in both Northern and Southern Hemispheres. The amount of water vapor in the atmosphere is of primary interest in meteorology. This amount present at any given time varies greatly due to temperature changes, evaporation from water surfaces, condensation, and precipitation (Cornish and Ives, 2009). When water vapor transforms from vapour to liquid water or ice, it releases a substantial amount of heat known as latent heat. Latent heat is a significant source of energy in the atmosphere, particularly during storms like thunderstorms and hurricanes (Ahrens, 2018). Furthermore, because it absorbs a large percentage of the earth's outgoing radiant radiation, water vapor is a powerful GHG. As a consequence, it is important in the earth's heat energy balance (Halo, 2016; Mahongo, 2016).

The atmosphere plays a major role in many processes affecting life on Earth. This includes the creation of different thermal zones over the earth; the variation in air temperature distribution and unequal heating that results in changes in atmospheric pressure and in turn initiation of the global wind system. Moreover, the formation of clouds, rain and snow. Furthermore, atmosphere shields life forms from dangerous solar radiation while also maintaining the gases that keep the planet alive.

1.1.1. The vertical structure of the atmosphere

The atmosphere has a mean mass of around 5.148 x 10^{18} kg, with three-quarters of it within 11 Km of the surface (Ahrens, 2018). With rising altitude, the atmosphere grows thinner, and there is no defined barrier between the atmosphere and outer space (Cornish and Ives, 2009). At an altitude of roughly 120 km, atmospheric effects become visible during atmospheric re-entry of spacecraft (Ahrens, 2018). Based on features such as temperature and composition, the atmosphere can be divided into several layers. The vertical layers composing our atmosphere are shown in Figure (1.2). This comprises (Ahrens, 2018; Wells, 2011):

1. Troposphere: is the earth's initial and lowest layer of atmosphere, containing 75% of the planetary atmosphere's total mass, 99% of the overall mass of water vapor and aerosols, and where most meteorological phenomena occur. It extends vertically up to 18-20 km above the mean sea level (MSL). This layer is characterized by a general drop in the air temperature, with a rate of 1 °C/km. The

temperature ranges from +15 °C to -56 °C. The troposphere ends with the tropopause separating it from the stratosphere. At all latitudes, the tropopause rises in the summer and falls in the winter.

2. Stratosphere: is the second atmospheric layer, which extends up to 50 km above the MSL. There is 1000 times less H_2O and 1000 times more O_3 in the stratosphere than in the troposphere. The temperature rises with altitude in this region. Heat is produced during the formation of O_3, and this heat is responsible for temperature increase ranging from an average of -51 °C at the tropopause to a maximum of about -15 °C at the stratosphere's top. The stratosphere is also known to be a cloud-free layer. Because there is little turbulence, most aircraft fly above the clouds and through the stratosphere. This layer is home to the ozone layer that protects life from the sun's ultraviolet radiation. This layer ends with the stratopause.
3. Mesosphere: This layer extends from 50 km above the MSL to about 85 km. As one ascends, the gases, including O_2 molecules, become lighter. As a result, the temperature decreases as one ascends, reaching around -90 °C near the top-edge of this layer, known as mesopause.
4. Thermosphere: This is known as the upper atmosphere and extends between 85 km and 690 km above the MSL.
5. Exosphere: above 690 km above the MSL, where atoms and molecules shoot off into space is the exosphere, which represents the top upper limit of our atmosphere.

In contrast to variant thermal behaviour at different atmospheric layers and altitudes, atmospheric pressure always falls as we rise in altitude: it falls swiftly at beginning, then more slowly at higher levels, similar to air density. The vertical thermal and pressure structures in the atmospheric layers are depicted in Figure (1.3).

It is worth mentioning that the topic of interest, air-sea interaction, can only be detected within the layer of the troposphere.

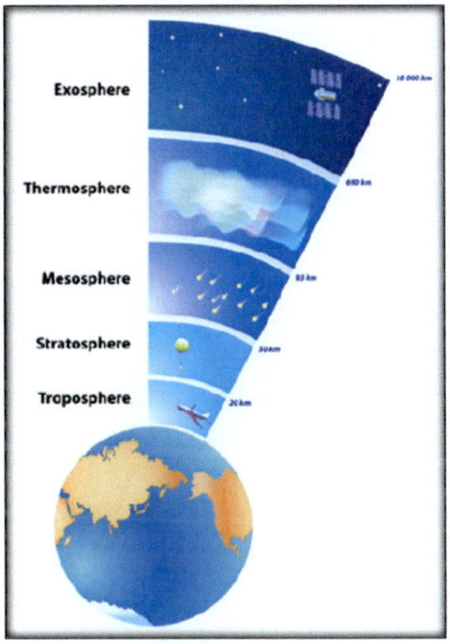

Figure 1.2. Vertical Structure of Atmosphere (Source: http://ete.cet.edu/gcc/?/volcanoes_layers/)

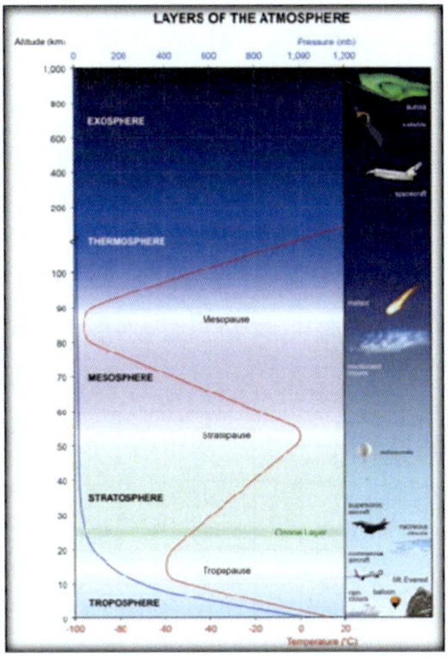

Figure 1.3. The vertical thermal and pressure structures in the atmosphere (Source: https://phys.org/news/2014-05-earth-magnetic-field-important-climate.html)

1.2. The Oceans

Oceans cover more than 70% of the earth's total surface area. They represent an interconnected water world. While the northern hemisphere is covered by 60% oceanic waters, ocean covers 80% of the southern hemisphere total surface area. The Pacific Ocean is the largest water body all over the globe representing 50.1% of the total oceanic surface areas, followed by the Atlantic Ocean (26%), followed by the Indian Ocean (20.5%) and followed by the Arctic Ocean (3.4%) (Pinet, 2009; Trujillo and Thurman, 2011). The Southern Ocean is mainly made of ice sheets and ice bergs (Pinet, 2009). It is important to briefly describe the main features of the global oceans; being the main regulators of the surrounding atmosphere.

1.2.1. The Pacific Ocean (Fig. 1.4) is the largest and oldest known ocean basin. This asymmetrical basin, which covers more than one-third of the earth's surface, is the world's greatest geographical feature (Trujillo and Thurman, 2011). It is characterized by the worldwide "Ring of Fire", i.e., it is geologically active by the existence of lots of trenches, island arcs, and volcanic mountain ranges. The Pacific Ocean is also featured by many sea mounts and guyot. The Pacific-Indian Ocean border runs from the Malay Peninsula to Sumatra, Java, Timor, Australia's Cape Londonderry, and Tasmania (Stewart, 2008).

1.2.2. The Atlantic Ocean (Fig. 1.4) is a symmetrical ocean, i.e., there is symmetry in its east-west continental margins, in its abyssal plains and in its Mid Ocean Ridge (MOR). Relatively, few trenches and island arcs (Caribbean) exist within the Atlantic basin. The meridian of Cape Agulhas marks the border between the Atlantic and Indian Oceans, while the line from Cape Horn to the South Shetland Islands represents the border with the Pacific Ocean (Stewart, 2008). The Atlantic Ocean is characterized by its adjacent (Gulf of Mexico) and marginal seas (Mediterranean Sea).

1.2.3. The Indian Ocean (Fig. 1.4) is to some extent smaller than the Atlantic Ocean, but it has almost the same average depth of 3840 m (Trujillo and Thurman, 2011). It is symmetrical about an "upside-down Y" created by a "triple junction". It is mostly in the southern hemisphere, formed due to Antarctic, African, Australian, Indian, and Asia plates movements (Pinet, 2009). Like the Atlantic, the Indian Ocean is featured by its adjacent seas, e.g. The Red Sea and Arabian Gulf.

1.2.4. The Arctic Ocean (Fig. 1.4) has the smallest ocean basin among the global system of oceans (~7% of the Pacific surface area) (Trujillo and Thurman, 2011). It is entirely land-locked except for its connection with the North Atlantic, and is characterized by very wide continental shelves.

1.2.5. The Antarctic Ocean (Fig. 1.4), oceanographers have recognized an additional ocean close to the continent of Antarctica in the southern hemisphere, south of latitude 50 °S and encircling Antarctica (Trujillo and Thurman, 2011). This has been named the Antarctic Ocean or the Southern Ocean. The Southern Ocean has experienced fast climate changes over the last 30 years, resulting in alterations in the marine ecology (Constable et al., 2014).

Figure 1.4. World Ocean Map
(Source: https://www.britannica.com/story/just-how-many-oceans-are-there)

The above-mentioned oceans, with their adjacent and connected seas, absorb and hold significantly more solar energy than land and atmosphere due to the large heat capacity of water (Halo, 2016). Consequently, oceans are able to change the overlying atmospheric conditions and weather. What happens is that the upper ocean connects surface forces from winds, heat, and fresh water to the quiescent deeper ocean, where this heat and fresh water are stored and released over longer time and planetary scales (Sprintall and Cronin, 2009; Sutherland et al., 2014). The released outgoing longwave radiation and latent heat flux return over half of the solar energy absorbed at the sea surface back to the atmosphere (Halo, 2016). Latent heat flux produces atmospheric vapor pressure, which is a key player in the earth's energy balance and drives important processes within the earth's hydrological cycle and diverse air-sea interaction processes (Talley et al., 2011). According to Bengtsson (2010), the total volume of accessible water on Earth's surface is around 1.5×10^9 km^3, of which 1.4×10^9 km^3 is oceanic, i.e. oceans retain more than 90% of the available water on Earth. Land ice and glaciers contain about 29×10^6 km^3 of water, while groundwater contains about 15×10^6 km^3 (Bengtsson, 2010). This reveals that the global hydrological cycle and the processes of interaction with the overlying layer of the atmosphere will interact differently among oceans and land masses over the earth. Also, differences will exist in the interaction process among the global water bodies of oceans and seas.

It is important not to forget that any interaction process between the ocean and the overlying atmosphere is conducted only in the surface layer of ocean not in the deeper layers. This surface layer comprises both an upper mixed layer exposed to direct atmospheric impact and a highly stratified zone below the mixed layer with severe vertical property gradients (Sprintall and Cronin, 2009), i.e. the surface and the cline layers (Fig. 1.5). The depth of this layer may vary from a few meters to a several hundred of meters depending on a density threshold relative to a reference depth (Sutherland et al., 2014).

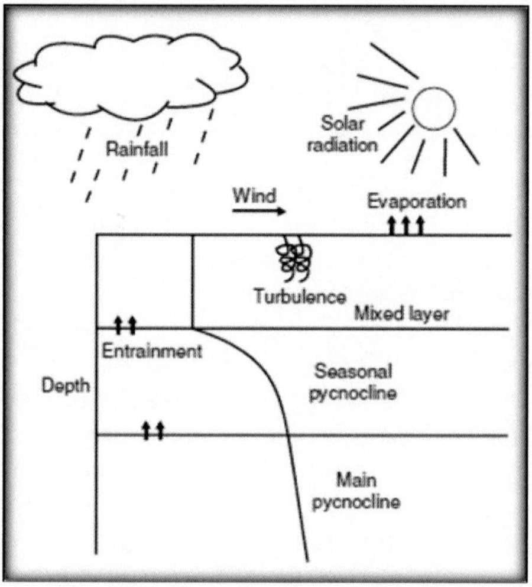

Figure 1.5. The vertical structure in the surface layer, as well as the forces and physics that regulate its existence. The mixed layer depth, the seasonal pycnocline, and the main pycnocline are all shown (Source: Sprintall and Cronin, 2009)

The two mentioned systems: atmosphere (1.1) and oceans (1.2) are continuously interacting and affecting each other; regulating the climate of the earth, maintaining its global cycles and impacting the anthropogenic activities. This interaction with its different and various processes, especially the abnormal ones, has motivated the performance of the present work to address the subject of 'Extreme Air-Sea Interaction with a Special Attention to the Mediterranean Sea and Red Sea'.

1.3. Aim and Objectives

This work aims at highlighting the processes induced by the air-sea interaction and the possible associated extreme vents, and to feature this within the basins of the Mediterranean and Red Sea.
The objectives of this work are: (1) To declare the importance of the air-sea interaction through its different processes, and the available tools to assess it, (2) To introduce the meaning of an extreme event in general and in air-sea interaction in particular, and (3) To spotlight extreme events in air-sea interaction in the basins of the Mediterranean and Red Sea, taking Marine Heatwaves (MHWs) as exemplar case studies.

This book consists of eight chapters, including this introductory one. The second chapter examines the air-sea interaction, including its importance, processes and quantification methodologies. Chapters three and four, respectively, focus on extreme events and MHWs. While chapter five shows information about the Mediterranean Sea, including examples of extreme events, using MHWs as a case study, chapter six shows the same information about the Red Sea. Conclusions and recommendations, and future perspectives are given in chapters seven and eight, respectively.

CHAPTER TWO

AIR-SEA INTERACTION PROCESSES

2.1. Definition of air-sea interaction

The air-sea interaction is one of the earliest subjects that attracted researchers in oceanography and meteorology for investigation and examination. This is dated back to the late 1940s, e.g. Schmidt (1947), who examined the sea-land breeze process, and early 1950s, e.g. Charnock (1951), who pointed out that despite the fact that the energy carried by the wind to the ocean is a tiny fraction of the radiation absorbed at the surface, wind-driven currents substantially dictate the locations where ocean energy is recycled back into the atmosphere, which establishes the cloudiness pattern.

According to Rogers (1995); Csanady (2004); Sprintall and Cronin (2009); Janssen et al., (2013); Mahongo (2016); Ren et al. (2020); Sauvage et al. (2020) and others, the air-sea interaction process can be defined as a two-way energy transfer mechanism, in which energy is transferred in one way from the atmosphere to the ocean surface mixed layer, and in turn energy is fed back from the ocean surface to the atmosphere. While the first mechanism affects the surface hydrography of oceans and drives the ocean circulation, the second drives variabilities in weather and climate (Rogers, 1995; Csanady, 2004; Trujillo and Thurman, 2011).

2.2. Importance of air-sea interaction

The air-sea interaction is a large field of research with significant progress made in many areas, e.g. El Nino, the circulation of oceans, the hydrological cycle, the heat fluxes, global climate change, etc.

The importance of the subject stems from the effect this interaction has on both the atmosphere and oceans on both small and large scales. The former comprises the heat exchange, evaporation rates and exchange of trace elements across the air sea interface and other important dynamical movement such as surges. The latter, on the other hand, comprises global climate and global circulation systems, e.g. winds and ocean surface circulation belts. The small-scale processes are important to solve global scale problems and to set boundaries and parameterization in ocean and climate general circulation models (Rogers, 1995). Generally speaking, the importance of the air-sea interaction processes can be pointed out as follow (Butterfield et al., 2005; Oliver, 2005; Suthers et al., 2011; Trujillo and Thurman, 2011; Christakos et al., 2014, 2016; Mahongo, 2016):

1. The major importance is the impact of the air-sea interaction processes on weather and climate because the global climate is determined by the mean exchange of heat, and anomalies in this exchange result in differences in local weather patterns.
2. The water exchange supplies almost the third of precipitation amount over the land, with a global mean surplus of evaporation over precipitation of 36×10^{15} kg/yr.
3. The exchange of climatically important gases, e.g. hydrogen sulphide (H_2S), nitrous oxide (N_2O), methane (CH_4), methyl iodide (CH_3I), carbon monoxide (CO), many of which play an important role in the greenhouse effect; influences the global climate on a long-term scale.
4. With phytoplankton in the oceans producing about half of the world's oxygen and releasing it into the atmosphere, the oceans operate as massive CO_2 sinks.
5. Oceans are an important supply of salt particles, which serve as a natural seeding source for cloud coalescence processes.
6. The fundamental cause of ocean circulation, particularly surface currents, is the exchange of momentum between the atmosphere and the ocean surface through wind stress.
7. Observed variations in sea surface temperature (SST) worldwide are mainly attributed to the air-sea interaction processes. The SST, in consequence, is the key player in the observed steric sea-level variations
8. Drought, heavy rain, heat waves, rogue waves, storm surges, and other observable extremes in atmospheric weather and oceanic status, are all linked to air-sea interaction and changes in atmospheric and oceanic conditions.
9. A revised modelling of air-sea interaction will contribute to the design of offshore wind farms using floating wind turbines, as the global interest in renewable energy grows.
10. Forecasting air-sea interaction processes also offers a better understanding of their nature and provides early warnings about damages caused by associated extreme events.

11. Ocean warming due to air-sea interaction processes will stress species by causing thermic changes in their environmental envelopes as well as increasing interspecies competition. These shifts become even more significant in shelf seas. Changes in coastal currents, for example, affect everything from primary production to fisheries productivity in south-eastern Australia.

The fundamental processes connecting the two domains: atmosphere and oceans are the energy input to the ocean by wind, the net freshwater flux and net surface heat flux. So, how does atmosphere affect oceans and how does the ocean, in turn, affect the overlying atmosphere? The answer can be briefly outlined as follow:

Variations in the air circulation have an impact on ocean circulation. The force of the wind blowing on the water surface exerts frictional drag on the ocean's surface layer and drives the surface currents in oceans and seas. The wind also mixes surface waters, forming the mixed layer, which has little vertical temperature variation. The thermocline, a narrow zone of rapidly declining temperature, lies underneath the mixed layer. The density of the water, which is affected by temperature and salinity, drives the circulation in the deep ocean below the thermocline (thermohaline circulation). Heat is stored in the deep ocean and released back into contact with the atmosphere thanks to vertical motions through and below the thermocline. In return, Oceans change atmospheric conditions and weather because of their ability to hold large amounts of heat and moisture. Tropical storms, for example, initially formed over warm ocean waters, provide the energy that allows hurricanes and typhoons to expand and move, often over land. Winter storms that deliver rain to the western United States originate over the North Pacific Ocean. Upwelling provides a cold differential in air temperature over the ocean and land in many coastal places, such as California, which is conducive to regular summer fog.

2.3. Examples of air-sea interaction processes

Before we go through the different types of fluxes and processes featuring the air-sea interaction phenomenon, it is worthwhile to start by addressing the mechanism of initiation (Stewart, 2008; Trujillo and Thurman, 2011; Wells, 2011) of such dynamical processes and movements.

Sun rays do not hit the earth equally. Between 35° and 40° north (N) and south (S) the equator, the sunlight strikes the earth at high angles and more energy is absorbed that is reflected back to the space. However, close to the poles the sunlight strikes the earth at lower angles, and with the abundance of a large ice cover energy reflected back to space is higher than that absorbed (Fig. 2.1). This is mainly attributed to the declination of the earth's axis with respect to the sun with an angle of 23.5°. The earth spins around its axis every 24 hours and in an elliptic orbit around the sun every 365 ¼ days. Therefore, with this tilt of rotation axis, the earth receives uneven heat amount gained by the oceans, and seasons develop over the year round (Fig. 2.2).

This imbalance of heat creates high and low pressure zones and, consequently, results in convention and moving air (winds) and water around the globe. The Coriolis force develops as the earth rotates, deflecting winds and surface currents to the right in Northern Hemisphere and to the left in Southern one. This way the global surface wind and surface current systems develop and dynamical movement of air-sea interaction is set.

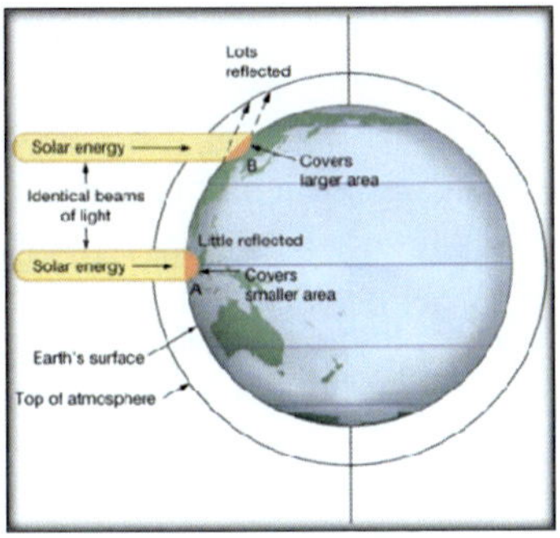

Figure 2.1. Uneven solar radiation on Earth (Source: Trujillo and Thurman, 2011)

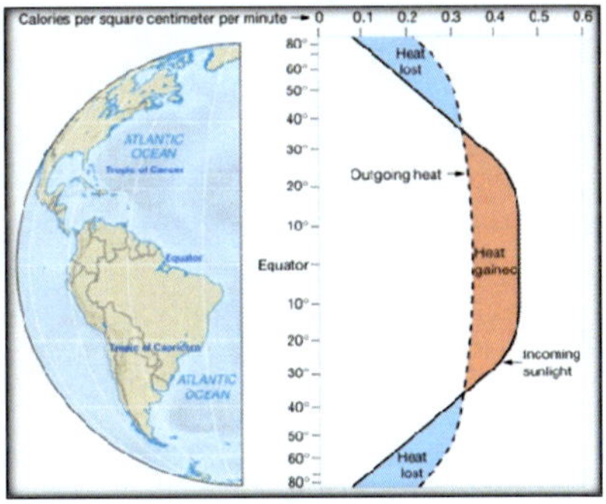

Figure 2.2. Heat gained and lost from the global ocean system (Source: Trujillo and Thurman, 2011)

Once in movement, the wind at the lower atmospheric boundary develops primitive air-sea interaction processes on local scale, e.g. sea and land breezes, sea fog and radiation fog. On regional scales, other processes develop influenced by the interaction between the sea surface and the overlying atmosphere, e.g. at mid latitudes storm systems featured by low pressure and warm/cold fronts occur; tropical cyclones, known in pacific and Atlantic regions as hurricanes, develop over tropical oceans where warm water results in moisty warm rising air to develop and by the sufficient effect of the Coriolis force rotation is enhanced and effects of cyclone appear.

Of course, on both local and regional scales, there are more air-sea interaction processes that can be detected and investigated. These depend not only on the variations in pressure and wind schemes solely, but also on differences in temperatures, surface ocean wave actions, sea-level variations and amounts of exchangeable gases. These processes will be discussed later.

The observed, investigated and simulated processes of air-sea interaction are termed fluxes (Ye et al., 2020). According to the examined process, these fluxes are named and categorized in four main types:

I. Momentum flux: this is the stress of the winds acting on the sea surface. This results in friction between the sea surface and atmosphere, which is a key component in air-sea interaction (He et al., 2018). On one hand, wind stress exports mechanical energy and participates in the turbulent exchange of heat and moisture in the atmosphere. On the other hand, wind stress drives ocean movement from the oceanic side (Ye et al., 2020). This stress sets up the ocean gyres and current systems, which in turn, redistribute heat and hydrographic properties in oceans.
II. Moisture flux: the source of water that sustains life on the planet is moisture fluxes over the ocean, i.e. evaporation and precipitation. Without the presence of water vapor in the atmosphere, the earth would be significantly colder and life as we know it would not exist (Stephens et al., 2020). These moisture fluxes are connected to the air-sea heat fluxes through the latent heat of evaporation. The ocean influences the atmosphere primarily through certain heat fluxes at the interaction boundary. This comprises the shortwave radiation flux (insolation), the net longwave flux (heat output from the sea surface towards the atmosphere), and the sensible and latent heat fluxes, which also usually transfer heat from sea to air. The former is the transfer of heat caused by the difference in temperature between the sea surface and the overlying air, while the latter is the heat absorbed on vaporization of the water (Csanady, 2004; Cronin et al., 2019). The latent heat is released to warm the atmosphere when the vapor condenses to form clouds (Cornish and Ives, 2009; Cronin et al., 2019).
III. Energy flux: is the amount of energy transferred from the blowing wind to the sea surface to create and influence surface ocean waves (He and Xu, 2016; He et al., 2018). Growing ocean waves have an impact on air-sea momentum and heat transmission, whereas

breaking ocean waves have an impact on upper ocean mixing processes (Janssen et al., 2013).

IV. Gas flux: CO_2, N_2O, and CH_4 exchange between the oceans and atmosphere are a crucial component of the climate system, and the oceans' tendency to absorb and desorb these gases alters both spatially and temporally (Holding et al., 2019).

As shown, air-sea interaction processes are diverse. Therefore, only selected examples will be discussed hereafter. Examples are downscaled from global to regional to local.

2.3.1. Examples of global scale air-sea interaction processes

By global scale air-sea interaction processes we mean these processes which occur all over the earth and their impacts extend to be sensed all over the globe. This can be presented by global wind systems, the global surface current system and its gyres, the balance of the different criteria such as heat, water and salt, etc.

2.3.1.1. Global wind system

The global wind system, also known as the global circulation system, is the first air-sea interaction process affecting the globe. The observed wind systems with their annual and seasonal variations comprise the global atmospheric circulation, which is the primary factor in determining the distribution of climatic zones, whereas variations in atmospheric and oceanic circulations are responsible for many of the observed longer-term fluctuations in climate (Ahrens, 2018). Inequalities in radiation distribution over the earth's surface and the earth's rotation are the two major controls on global wind circulation. The global atmospheric circulation is driven by global radiation distribution and gravity, while the earth's rotation determines its shape (Oliver, 2005). In this global circulation system, the atmosphere transports warm air towards the poles and the cold air towards the equator (Cornish and Ives, 2009; Ahrens, 2018). Although it appears to be simple, the actual air flow is complex and not everything is known about it to date (Ahrens, 2018).

This circulation system was first examined by George Hadley who made three assumptions to describe the global air flow (Oliver, 2005; Stewart, 2008; Trujillo and Thurman, 2011: (1) the earth is totally covered with water with no land mass, so that there is no thermal variation on the earth surface. (2) Sun is 100% normal to the earth so that there is no seasonality and (3) Earth does not rotate, so that the Coriolis force is ignored and the only acting force deriving the motion is the pressure gradient. Accordingly, Hadley got the atmospheric cell flow named after him, and his theory is named the single cell model theory (Fig. 2.3).

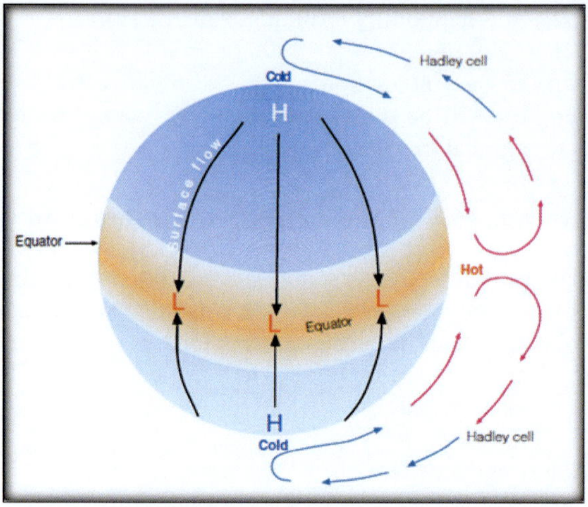

Figure 2.3. Hadley's single cell model describing the global surface air flow (Source: Oliver, 2005)

As the earth is actually rotating, different pressure zones develop from the equator poleward, and as air flows its pathway is deflected to the right in the northern hemisphere and to the left in the southern hemisphere affected by the Coriolis force. This pattern of deviation has led to what is known as the three cell wind system, which describes the global surface wind system. This system (Fig. 2.4) consists of three cells as follow:

(1) Hadley Cell: developed at the equator, where the air is warm over equatorial waters, horizontal pressure gradients are weak, and winds are light. The 'doldrums' are a term used to describe this area. This low pressure zone extends to 30° N/S where air becomes denser and sinks to flow equatorward. Winds in this zone are known as 'Tarde Winds'
(2) Ferrell Cell: Not all surface air moves equatorward at latitude 30° N/S. Some air flows towards the poles and deflects toward the east, resulting in a more or less westerly air flow in both hemispheres,

known as 'The Westerlies'. This forms the Ferrell cell and extends to 60° latitude in the two hemispheres.

(3) Polar Cell: The mild air from the Ferrell cell at 60° N/S encounters cold air moving down from the poles. These two air masses with contrasting temperatures do not mix easily. They are separated by the polar front, a zone of low pressure—the subpolar low—where surface air converges and rises, causing storms to form, known as 'Polar Easterlies'.

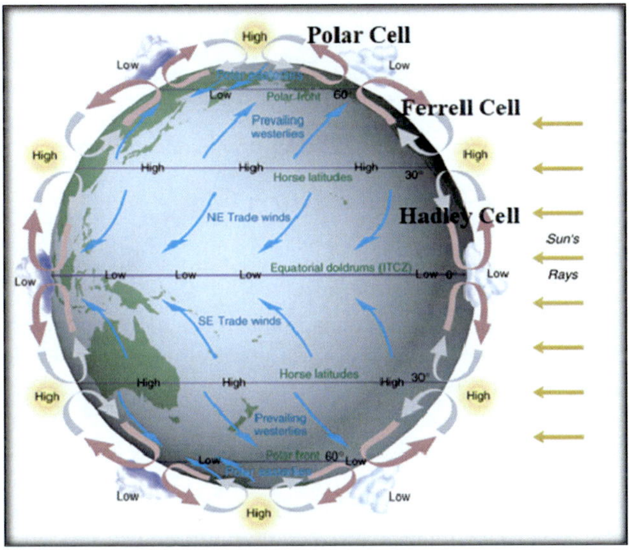

Figure 2.4. The three global cells circulation system with nomenclatures of dominant winds and circulation cells (Adapted from Trujillo and Thurman, 2011)

Rain and wet environments are created by rising air and low pressure at 0° and 60° latitudes, while high pressure, sinking air creates drier conditions at 30° and 90° latitudes (Fig. 2.5).

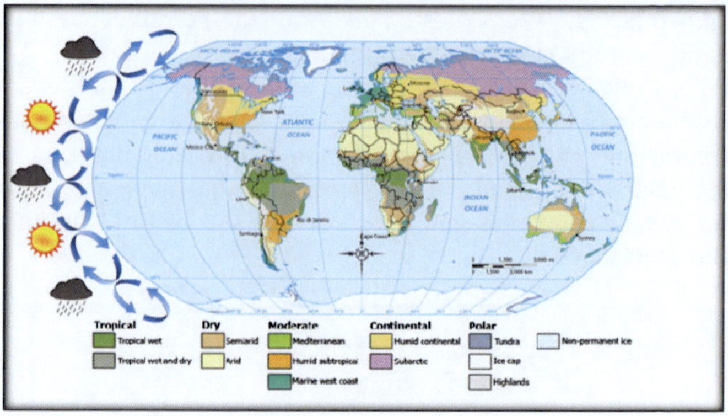

Figure 2.5. Major climatic regions of the world in relation to atmospheric convection cells (Source: Webb, 2021)

2.3.1.2. Global surface circulation system

A current is a large stream of water that flows in a regular pattern through the ocean (Pinet, 2009). By transporting heat, carbon, oxygen, and nutrients throughout the world's ocean, ocean circulation plays a critical role in regulating climate and supporting marine life (Talley et al., 2011). Oceanographers categorize ocean currents in two categories: surface and deep currents. Generally, currents in the upper layers of the ocean are caused by wind stress, while those below are caused by the distributions of temperature and salinity, but this distinction is often blurred, and the two are linked by vertical water movements (Pinet, 2009). What belongs to the present work is the global surface circulation system as one on the global air-sea interaction processes. According to the global wind system (section 2.3.1.1) the prevailing winds, predominantly easterlies in the tropics and westerlies in the mid-latitudes, drive the wind-driven circulation. Surprisingly, merely about 2% of the wind energy is transmitted to the ocean surface, so a 25 m/s wind will create only a 0.5 m/s current (Trujillo and Thurman, 2011).

The upper ocean moves in a balance of frictional and Coriolis forces, known as Ekman transport as the winds blow above the ocean surface. This mechanism empowers the perpendicular net Ekman mass transport (Fig. 2.6) to the affecting wind (to the right and left in Northern Hemisphere and Southern Hemisphere, respectively). This transport causes divergence and convergence, which cause upwelling (the upward movement of interior waters) and down-welling, i.e., sinking of surface waters (Morrison et al., 2015).

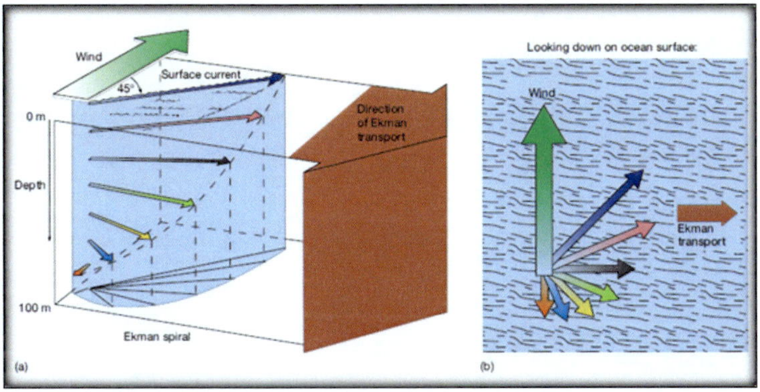

Figure 2.6. Ekman spiral and Ekman transport in perspective (a) and top view (b) (Source: Trujillo and Thurman, 2011)

Surface currents would generally follow the world's major wind belts if there were no continents. Because of the trade winds, a current would flow between 0-30° latitudes in each hemisphere, a second current would flow between 30-60° latitudes as a result of the prevailing westerlies, and a third current would flow between 60-90° latitudes as a result of the polar easterlies (Trujillo and Thurman, 2011). However, in the real world, beside the Coriolis Effect, ocean surface currents are driven by more than just the global wind belts. Indeed, the nature and direction of the surface current flow in each ocean basin is influenced by the distribution of continents on Earth (Pinet, 2009; Trujillo and Thurman, 2011). As a consequence, the subtropical and subpolar gyres are formed by a series of zonal currents (e.g. North Equatorial and South Equatorial currents), cold surface currents (e.g. California and Chile/Peru Currents), and warm surface currents (e.g. Kuroshio Current and Gulf Stream) as depicted in Figure (2.7). The subtropical gyres are particularly important because they convey heat from the equator to the poles via warm surface currents and provide oxygen to low-latitude interior waters via subsurface return flow of these surface waters (Talley et al., 2011).

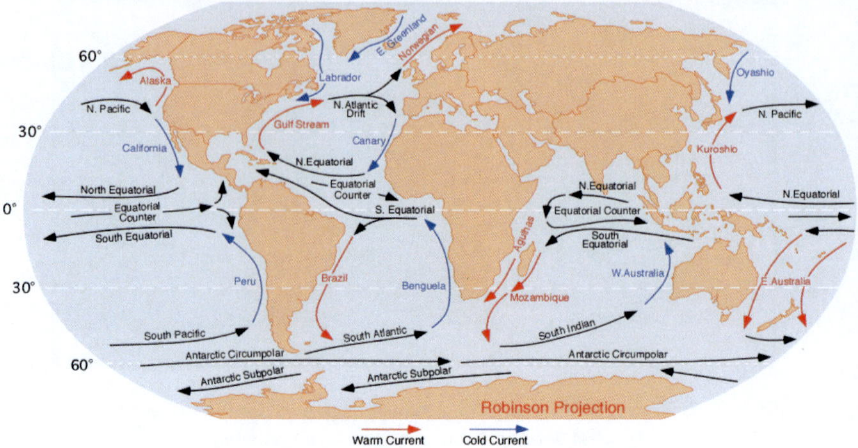

Figure 2.7. Global surface ocean currents
(Source: http://oceanmotion.org/global-surface-currents.htm)

2.3.1.3. Global Heat Budget

As mentioned in the previous section, surface currents, cold and warm, are responsible for heat transfer around the globe. Practically, turbulent and radiative interactions of heat between the ocean surface and the overlying adjacent layer of atmosphere, known as heat fluxes, with ocean surface wind stress and currents are crucial climate variables influencing weather and climate (Cronin et al., 2019). Therefore, the examination of the global heat balance is one of the vital topics examined in the studies of the air-sea interaction process.

The global heat budget is the balance of incoming and outgoing heat as processed between the Sun, atmosphere, land and oceans (Webb, 2021). To maintain constant conditions, the budget must be balanced so that the incoming heat equals the outgoing heat, just like any other budget. 30% of the solar radiation that reaches Earth is reflected back into space, while 70% is absorbed by the earth (47%) and atmosphere (23%) respectively (Trujillo and Thurman, 2011; Webb, 2021). Because of its large surface area (~71% of the earth's surface) and low surface albedo, the ocean absorbs the greatest fraction of solar radiation incident on the earth's surface (Wells, 2011). The different components affecting the global heat budget are shown in Figure (2.8).

From an interaction perspective, the lower atmosphere is heated (and cooled) by the oceans which impacts on weather and climate. The ocean surface cools as seawater evaporates, and when the moisture condenses into cloud droplets, this heat is released, warming the atmosphere. This moistening and warming causes the air to become buoyant, causing low-level baroclinicity and atmospheric convection, resulting in wind convergence at the surface and divergence skyward (Cronin et al., 2019).

The mechanism behind the global heat budget depends on specified interactive processes, the first of which is the insolation (Incoming Solar Radiation). The significant other terms in the budget at the surface of the ocean are 1. the net infra-red radiation from the sea; 2. the sensible heat flux which is the flux of heat out from the sea due to conduction; 3. the latent heat flux, which is the flux of heat carried by evaporated water; and 4. the heat transferred by currents (Stewart, 2008; Wells, 2011; Webb, 2021).

Insolation:

The latitude, season, time of day, and cloudiness are the primary determinants of incoming solar radiation (Stewart, 2008). Short-wave radiation from the sun penetrates the surface of the ocean to depths of tens of meters, heating the ocean mixed layer at rates that exceed other components of the heat budget around mid-day under clear skies (Csanady, 2004). In different locations and seasons, different processes dominate the heat budget of the oceanic mixed layer, resulting in a complex geographical distribution of annual mean oceanic heat loss and heat gain (Csanady, 2004). The way this radiation is absorbed is crucial to understanding both the thermal distribution in the ocean and the ocean's stability. Furthermore, the biological activity of the ocean is essentially dependent on microscopic phytoplankton's direct absorption of incident solar radiation (Wells, 2011). The transfer of heat across or through the air-sea interface is referred to as a heat flux (Stewart, 2008). The energy that the earth absorbs returns to the atmosphere via three processes: conduction, radiation, and latent heat (phase change) (Webb, 2021).

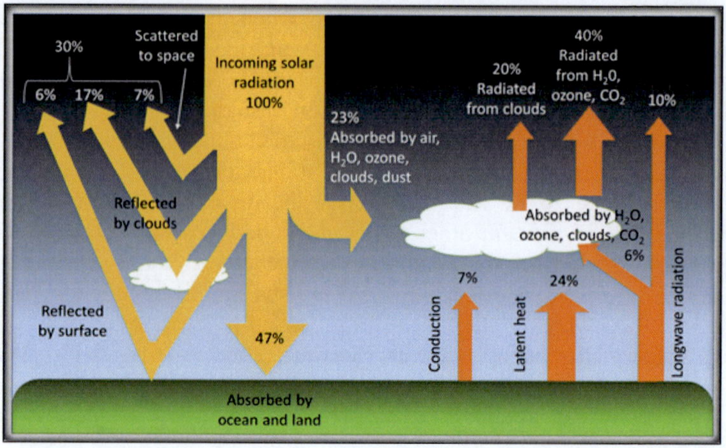

Figure 2.8. Global heat budget (Source: Webb, 2021)

Infra-red Radiation:

According to Planck's equation, the sea surface radiates like a blackbody at wavelengths close to 10 μm. These radiated wavelengths are strongly absorbed by clouds, and somewhat by water vapor. Therefore, the infra-red back radiation from the sea surface to atmosphere is dependent on (Stewart, 2008): 1. Cloud heights and thickness, 2. Atmospheric water vapor content, 3. SST and 4. Ice/snow cover.

Conduction:

This is the direct transfer of heat between the surface and the atmosphere. Because air is a poor thermal conductor, conduction accounts for only a small portion of the energy transfer between the earth and the atmosphere, accounting for about 7% of the incoming solar energy (Webb, 2021). In the field of air-sea interaction research, conduction process is referred to as the sensible heat flux. Wind speed and air-sea temperature difference affect the sensible heat flux. High fluxes are caused by strong winds and large temperature differences (Stewart, 2008).

Latent heat flux:

Latent heat flux is the energy lost/gained at the sea surface during evaporation/condensation as water moves from a higher to lower energy state (or vice versa) (Dugdale et al., 2017). Wind speed and relative humidity have the greatest influence on latent heat flux. High winds and dry air evaporate far more water than weak winds and high relative humidity (Stewart, 2008).

Advection:

This is the heat transferred, whether vertically or horizontally, from one location to another by ocean currents, which impact on the SST and in consequence on the calculated heat budget (He et al., 2017)

In a normal situation on a global scale, the sum of all mentioned heat fluxes must be zero. Otherwise, the ecosystem will heat up or cool down and therefore global climate changes, a problem mankind examines nowadays. These disturbances, in fact, cause teleconnections in the atmosphere, influencing weather and climate from afar (Cronin et al., 2019). For example, because the surface latent heat flux is a direct heat source for the atmosphere when water vapor condenses, precipitation is inextricably linked to the energy budget in the atmosphere (Myhre et al., 2018). Changes in atmospheric radiative cooling caused by shortwave or longwave radiation can occur as a direct result of a climate forcing mechanism, such as increased insolation or CO_2 concentration, or because of climate feedback caused by changes in temperature difference between the surface ocean and the overlying atmosphere (Myhre et al., 2018). These changes impact on the precipitation regime and would result in a total alteration for the known precipitation map over the globe. On a basin-wise scale, on the other hand, it is accepted to examine either gain (+ W/m^2) or loss (- W/m^2) in this heat budget, as shown in Figure (2.9) depicting the heat budget of the Atlantic Basin with a more than 100 W/m2 heat loss in the northern Atlantic basin and a more than 25 W/m^2 heat gain in the southern Atlantic basin.

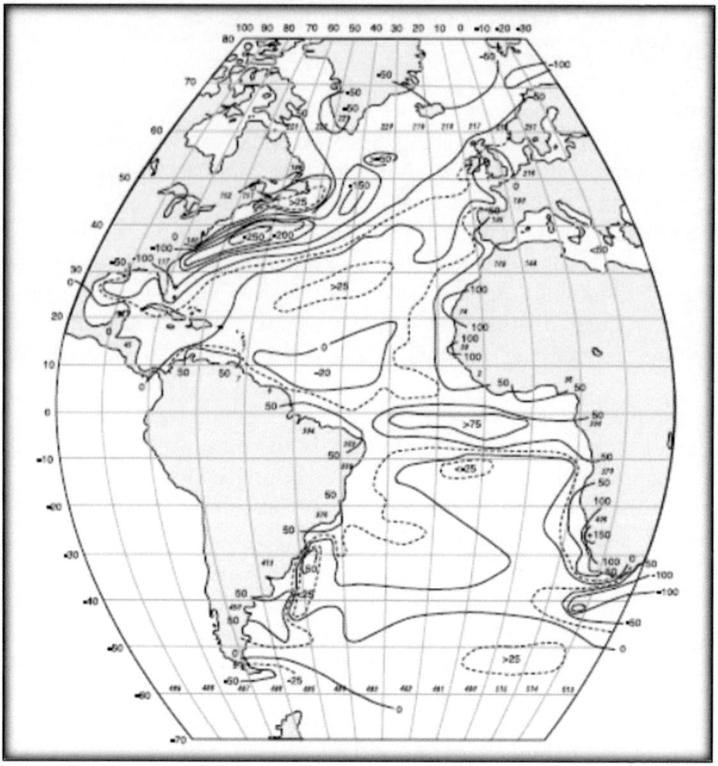

Figure 2.9. Map of the net annual heat budget in the Atlantic Ocean (Source: Csanady, 2004)

2.3.2. Examples of regional scale air-sea interaction processes

The above-mentioned global scale air-interaction processes in addition to many others are the principle drivers of many regional well-observed and well-examined air-sea interaction phenomena all over the globe. Hereafter, two major of these regional phenomena will be discussed: El Nino Southern Oscillations (ENSO) and to the Monsoon.

2.3.2.1. El Nino Southern Oscillations (ENSO)

The El Nino Southern Oscillation (ENSO) is the most effective interannual climatological phenomenon in the tropical ocean–atmosphere system (Collins, 2000; Lau and Waliser, 2012). ENSO entails significant changes in tropical convection, large-scale circulation, and weather patterns all over the world (Lau, 1985). Because of its power to alter global air circulation, ENSO is one of the most important climate phenomena on the planet, influencing temperature and precipitation schemes, and may result in severe extreme events according to its intensity (Pedreira Junior et al., 2020). Therefore, ENSO is believed to have a direct impact on human life and terrestrial and marine ecosystems around the world due to its global influence on the atmosphere and oceans (Chen et al., 2020). According to Ahrens (2018), ENSO can be defined as a tropical Pacific condition in which a reversal of surface air pressure at opposite ends of the Pacific Ocean induces westerly winds, a strengthening of the Equatorial Countercurrent, and widespread ocean warming. The main driving mechanism to induce ENSO is the irregular periodic variation in SST and winds over the tropical eastern Pacific Ocean (Collins, 2000). Extreme weather (such as floods and droughts) is caused by the oscillations of this climate pattern in many parts of the world. Developing countries reliant on agriculture and fishing, particularly those bordering the Pacific Ocean, are badly affected. Although research on ENSO started as early as the 1960s, e.g. Bjerknes (1969), the mechanisms which cause the oscillation are still being researched (Wang, 2018).

Though ENSO is a single climate phenomenon, it can exist in one of three states, or phases (Zebiak, 1993): Neutral, El Nino and La Nina. Because ENSO is a coupled climate phenomenon, the two opposing phases, ‘El Nino’ and ‘La Nina’, necessitate changes in both the ocean and atmosphere. The term "neutral" is located in the middle of the ENSO spectrum.

1. Neutral phase (Fig. 2.10): When the tropical Pacific Ocean is in a neutral state (neither El Nino nor La Nina), trade winds blow east to west across its surface, bringing warm moist air and warmer surface waters to the western Pacific, while keeping the central Pacific relatively cool. The west has a deeper thermocline than the east (Anonymus, 2012). ENSO conditions are described as neutral if the SST variation from the climatology is ≤ 0.5 °C. The transition between ENSO's warm (El Nino) and cold (La Nina) phases is defined as neutral conditions. During this phase, the ocean SST, tropical precipitation, and wind patterns are near average (Wang et al., 1999).
2. El Nino phase (Fig. 2.11): During an El Nino event, trade winds diminish or even invert, allowing warmer-than-normal water to flow into the tropical Pacific Ocean's central and eastern regions

(Wang et al., 1999; Wang, 2018). The deepening of the thermocline in the central to eastern Pacific is associated with these warmer-than-normal ocean temperatures. Warmer SST is also caused by a weaker upwelling of cooler ocean waters from below (Wang and Picaut, 2013).

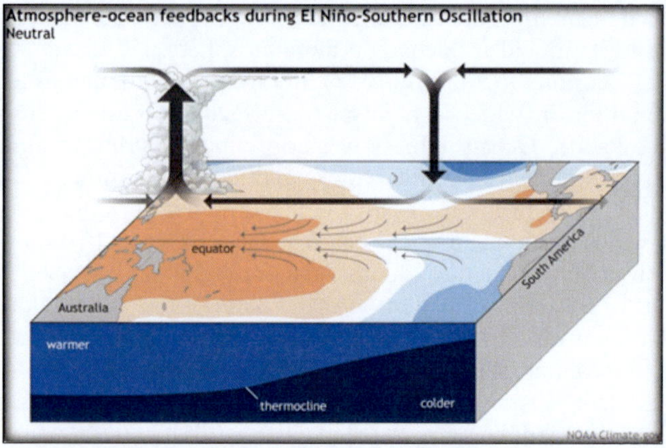

Figure 2.10. The ENSO neutral phase
(Source: https://www.climate.gov/sites/default/files/ENSOblog_Bjerknes-feedback-Neutral_large.jpg)

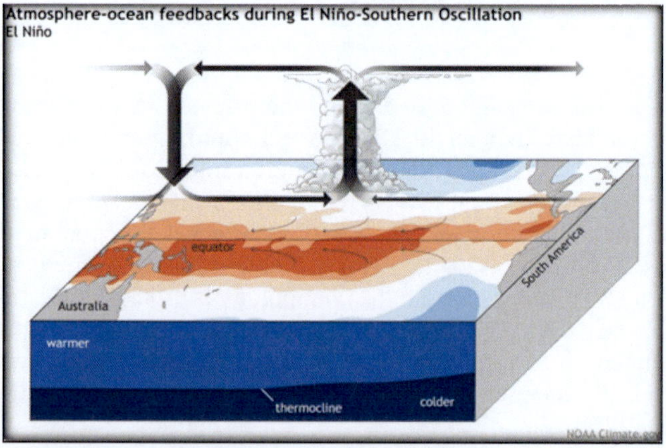

Figure 2.11. El Nino phase
(Source: https://www.climate.gov/sites/default/files/ENSOblog_Bjerknes-feedback-ElNino_large.jpg)

3. La Nina phase (Fig. 2.12): in this phase, the pool of warmer water is limited to the far western tropical Pacific as the trade winds accelerate, resulting in higher-than-normal SST north of Australia. The thermocline rises closer to the surface as upwelling strengthens and SST over the central and eastern tropical Pacific Ocean becomes cooler than usual. Cool waters from the deep ocean are brought to the surface when upwelling strengthens (Trenberth et al., 2007).

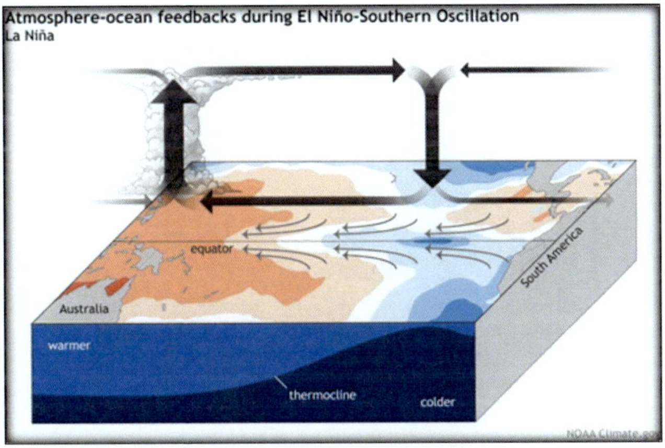

Figure 2.12. La Nina phase
(Source: https://www.climate.gov/sites/default/files/ENSOblog_Bjerknes-feedback-LaNina_large.jpg)

The associated variations in the SST anomalies (deviation from the mean) during the warm (El Nino) and cold (La Nina) phases of ENSO are depicted in Figure (2.13).

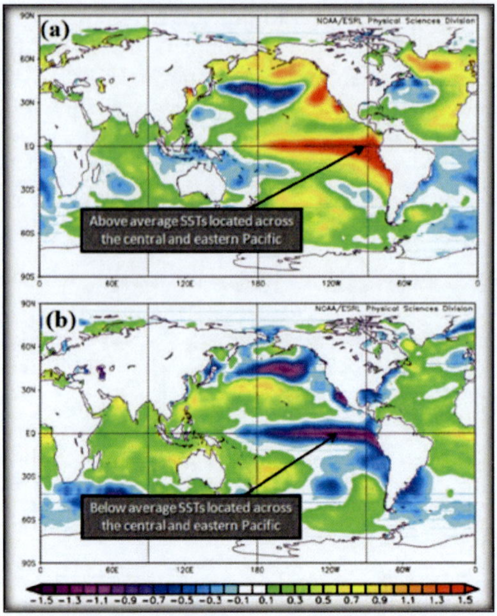

Figure 2.13. SST anomalies during (a) warm and (b) cold phases of ENSO (Source: Lawal, 2015)

2.3.2.2. Monsoon

A second large scale air-sea interaction phenomenon impacting the global climate, humans and various global ecosystems is the monsoon. The term monsoon comes from the Arabic word mausim, which means "season". Today, the term 'monsoon' is defined as to describe a wind system that changes direction seasonally, blowing from one direction in the summer and the opposite direction in the winter driven by the exchange of energy and moisture between the land, ocean, and atmosphere (Mölders and Kramm, 2014; Ahrens, 2018). Factually, the monsoon is a defining feature of Earth's climate, and its concept has evolved over time (Wang et al., 2011). Monsoons are a significant component of tropical and subtropical climates in many parts of the world, represented by wet summers and dry winters, as well as a seasonal reversal of the predominant winds (Geen et al., 2020). Most of the countries located between 35° N and 25° S and 30° W and 170° E (Fig. 2.14) meet the monsoon criteria and are thus widely accepted as being part of the world's most important monsoon domain (Oliver, 2005). However, the global monsoon system consists of seven major systems (Lau et al., 2007; Wang et al., 2011): East Asian Monsoon (EAM), South Asian Monsoon (SAM), African Monsoon (AM), Indian Monsoon (IM), Australian Monsoon (AM), North America Mexican Monsoon System (NAMS) and South America Monsoon System (SAMS). The annual cycle of solar radiation drives and synchronizes all seven regional monsoons, which are linked by the global divergent circulation (Wang et al., 2011).

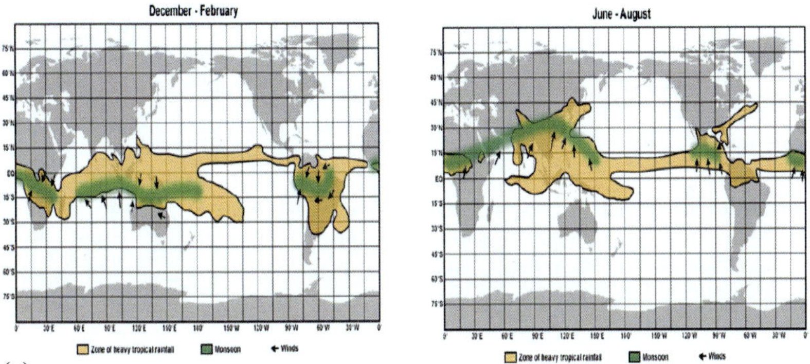

(a) (b)

Figure 2.14. Regions of Monsoon in (a) winter months and (b) summer months (Source: Rohli and Li, 2021)

Despite this global link between the seven systems, regional monsoons appear to act essentially as independent systems on inter-decadal to intra-seasonal time scales, with some degree of coordination by teleconnections to ENSO (Yim et al., 2014). Even within the same region, patterns of inter-annual variations of features associated with the monsoon may have distinct behaviour (Geen et al., 2020). Despite the fact that any of the seven monsoon events is accompanied by variations in wind, atmospheric pressure and SST, precipitation and rainfall pattern remain the most studied monsoon aspect (Oliver, 2005). This may be attributed to the fact that agricultural activities in the affected regions are sensitive to the inter-annual to intra-seasonal variations in the rainfall pattern associated with the monsoon. This is explained for the Asian case of different countries, e.g. China, India, Bangladesh, Sri Lanka (Yokoyama et al., 2020), Australia (Sudmeyer et al., 2016), West Africa (Froidurot and Diedhiou, 2017), and Mexico (Granados et al., 2017). In the present work, light is spot on the tropical monsoon affecting the Red Sea: the IM.

The Western Indian Ocean (WIO) has more significant seasonal wind reversals than the rest of the Indian Ocean, making it an important air–sea interaction zone (Mahongo, 2016). Indeed, it is the only place on the planet where reversing seasonal winds cause major ocean surface currents to reverse direction (Trujillo and Thurman, 2011). The northeast (NE) and southeast (SE) trade winds, which lead to the NE and SE monsoons, are two types of seasonal wind systems that have been identified, and are affected by the seasonal variations in the Inter-Tropical Convergence Zone (ITCZ) (Trujillo and Thurman, 2011; Mahongo, 2016). The geographical orientation of the Indian Ocean is the most important factor in determining the generation of the monsoon seasonal wind pattern (Mahongo, 2016).

The Indian Ocean is bordered to the north by the Asian continent, which has a significant impact on the region's climate. Another factor is the presence of the East African Highlands, which contribute significantly in establishing upper air flow, particularly the Somali Jet, which has a significant effect on WIO weather patterns (Slingo et al., 2005). As explained by Trujillo and Thurman (2011), offshore winds cause the North Equatorial Current to flow from E to W during the winter NE monsoon, and its extension, the Somali Current, to flow south along Africa's coast. There is also an Equatorial Countercurrent established. The winds reverse during the summer southwest (SW) monsoon, causing the North Equatorial Current to be replaced by the Southwest Monsoon Current, which flows in the opposite direction (Fig. 2.15). Apart from the oceanic circulation, during the summer SW monsoon approximately 80% of India's annual precipitation falls, providing water to 20 crops during prime agricultural season (Katzenberger et al., 2020). This heavy rainfall pattern is mainly attributed to the SST and moisture flux transport in the Indian Ocean in the preceding winter and spring to the summer monsoon (Katzenberger et al., 2020).

Research on the extended effect of IM on the Indian Ocean's marginal seas, e.g. the Arabian Gulf and the Red Sea, has been performed. This effect has been seen to cover not only surface circulation but also some other surface oceanographic features in addition to the weather in the region it reaches. Winds in the southern part of the Red Sea are influenced by the IM. The prevailing wind across the Arabian Sea during the NE monsoon (October to May) is north-easterly, turning to easterly in the Gulf of Aden and south-easterly near the Strait of Bab El- Mandab and the southern Red Sea. However, during the SW monsoon (June–September), the strongest winds are from the WSW over the Gulf of Aden and Arabian Sea, and from the NW over the southern Red Sea (Schott and McCreary, 2001; Aiki et al., 2006; Morcos and Abdallah, 2012). The monsoonal reversal

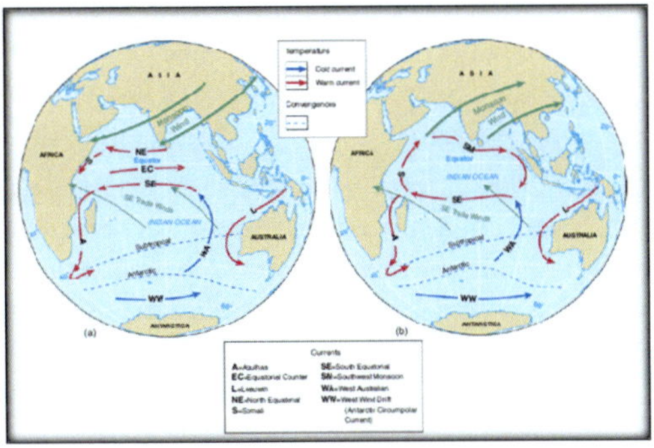

Figure 2.15. Seasonal surface currents in the Indian Ocean affected by (a) the winter NE monsoon and (b) summer SE monsoon (Source: Trujillo and Thurman, 2011)

of winds, combined with the thermohaline circulation, causes alterations in circulation in the Strait of Bab El-Mandab and Gulf of Aden, according to published literature, e.g. Sofianos and Johns (2003); Schmidt et al. (2020). During the summer, the SW monsoon's stronger winds push surface water to flow out of the Gulf of Aden and into the Arabian Sea, forming an anticyclonic gyre in the middle Gulf of Aden (Al Saafani and Shenoi, 2007). During the summer months of July and September, strong winds of the SW monsoon cause an intense seasonal upwelling over a large area off the coast of Southern Arabia (Al-Akhaly et al., 2020).

Due to a lack of riverine input and minimal precipitation, the Red Sea's nutrient-depleted ecosystem is primarily dependent on the horizontal intrusion of nutrient-rich waters from the Indian Ocean via the Gulf of Aden (Churchill et al., 2014; Triantafyllou et al., 2014). During the winter monsoon, the reversal of wind direction promotes the northward advection of nutrient-rich waters through into the Red Sea, resulting in the seasonal succession of Red Sea phytoplankton (Raitsos et al., 2013, 2015). Ionita et al. (2014) used coral oxygen isotopes from the southern Red Sea during the winter. They concluded that the temperature and salinity variations from 1960 to 1990 were caused by the IM. The sea level variability along the Red Sea is affected by the reversal system of the IM impacting on the Basin e.g. (Sultan et al., 1995; Sultan and Elghribi, 2003; Landerer et al., 2008; Manasrah et al., 2009; Stammer et al., 2013; Nidheesh et al., 2017; Alawad et al., 2019).

2.3.3. Examples of local scale air-sea interaction processes

On a local scale, many air-sea interaction processes can be observed, measured and calculated. These processes have impact on the anthropogenic activities in the neighborhood regions where they occur. Examples of such processes include, but not limited to, sea-land breezes, evaporation process, heat, water and salt budgets of aquatic basins, atmospheric surges, surface waves, etc. The records of the local meteorological parameters are key players in the investigation of so many of these processes. Hereafter, a few examples of local scale air-sea interaction processes will be briefly introduced.

2.3.3.1. Sea-land breezes

A breeze is a gentle wind that blows at a consistent speed. Sea-land breezes occur in or near vast bodies of water such as oceans, seas, and rivers. This is one of the oldest air-sea interaction mechanisms examined by researchers. Schmidt (1947) defined the phenomena as regular changes in onshore and offshore winds that occur in coastal areas during periods of good weather. This process is a distinct feature that influences local weather and climate, not just at the shore but also in neighboring inland areas. Both breezes are beneficial because they provide some relief from the hot and humid weather. Sea–land breezes are caused by daily differential warming between land and sea, which creates horizontal pressure gradients and allow cold and humid marine air to enter landwards (Abbs and Physick, 1992; Pokhrel and Lee, 2011; Naor et al., 2017). The differing responses of land and water surfaces to temperature fluctuations related to their distinct thermal characteristics and energy balance are principally responsible for the sea–land breeze circulation along coastlines (Abbs and Physick, 1992). In addition, variations in latitude, sea temperature, and topography affect the growth and decay of the breeze system in every coastal region on Earth (Yoshikado, 1981). The strength and direction of wind patterns, atmospheric stability, and local topography factors all influence sea–land breezes. Hence, breezes are thought to be a fairly local phenomenon (Carnero et al., 2005; Naor et al., 2017). A sea breeze develops during the day affected by the amount of land heating and blows onshore throughout the day (Yoshikado, 1981). In contrast, land breeze blows offshore at night (Abbs and Physick, 1992). Though sea and land breezes are more common in the spring and summer, they can occur at any time of year when there are enough light days (Azorin-Molina et al., 2011). Figure (2.16) is a schematic diagram showing the air flows during sea and land breezes at coastal areas.

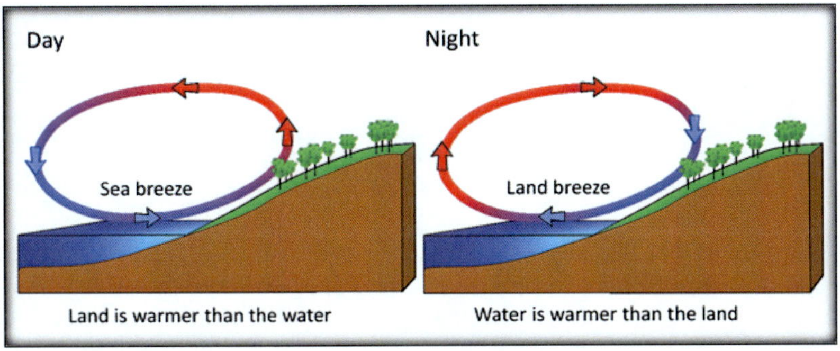

Figure 2.16. A schematic diagram of the sea and land breezes during day and night (Source: Webb, 2021)

Theoretical and observational research of sea breeze circulations have been performed in many regions worldwide: South-eastern Australia (Abbs and Physick, 1992); California (Banta et al., 1993); Chile (Kalthoff et al., 2002); Indonesia (Hadi et al. 2002); Adriatic Sea (Prtenjak, 2003); the Arabian Gulf (Eager et al., 2008); New Jersey coast (Seroka et al., 2018); Columbia (Perez et al., 2018); the Red Sea (Davis et al., 2019); South-west Australia (Rafiq et al., 2020) and the southern Levantine (El-Geziry et al., 2021).

Trials to identify sea breeze days comprise, but not limited to, the work of Furberg et al. (2002); Bigot and Planchon (2003); Dunsmuir et al. (2003); Azorin-Molina et al. (2011). Various criteria are used to determine such days: Since the diurnal reversal of wind direction from offshore to onshore was utilized as an identifier of sea breeze days, the characteristics of wind regime have been an important factor in many sea breeze studies, e.g. Furberg et al. (2002); Miller et al. (2003); Azorin-Molina and Martin-Vide (2007); Azorin-Molina and Chen (2009); El-Geziry et al. (2021). Also, the decrease in air temperature connected to the increase in relative humidity is another criteria used to specify a sea breeze day (Sumner, 1977; Abbs, 1986; Abbs and Physick, 1992). A sea breeze day can also be identified by the clarity of the sky, regardless of the presence of clouds. The cloudiness of more than 5 oktas is thought to lessen the likelihood of a sea breeze by slowing the land's heating process (Masouleh, 2015). To perfectly identify sea breeze days at any coastal location, a combination of these identifiers is highly recommended.

It is critical to understand the physical nature of sea–land breezes to understand their behavior and influence on coastal activities. While observational and modelling studies are required to understand the mechanisms that control it, the study of sea–land breeze climatology is an important complement to those studies. It is hoped that the knowledge gained will be useful to weather forecasters and those whose future plans rely on wind (El-Geziry et al., 2021).

2.3.3.2. Evaporation rates

Evaporation is the process of changing the surface seawater to vapor, and hence it is the main supplier of water vapor found in the atmosphere (Cornish and Ives, 2009). Understanding atmospheric moisture is especially important in the coastal zones and in air-sea interaction processes. In the tropical seas, the greatest water vapor content is roughly 4% of the atmosphere's mass, whereas near the poles in winter, it's virtually zero (Rohli and Li, 2021). This is schematically represented in Figure (2.17).

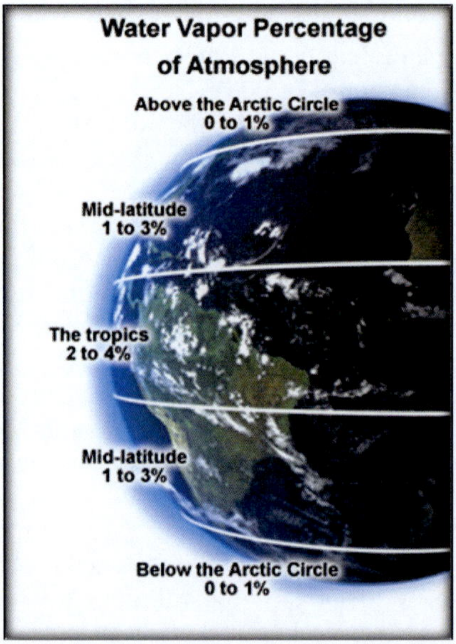

Figure 2.17. The distribution of latitudinal surface water vapor percentage (Source: Rohli and Li, 2021)

In oceans, about 90% of the available heat is used to evaporate water (Wells, 2011). According to Stewart (2008); Trujillo and Thurman (2011) and Webb (2021), oceans and seas account for 86% of global evaporation.

The latent heat of evaporation of water maintains the thermal equilibrium of the atmosphere (Wells, 2011). Heat lost from the sea surface to the atmosphere via the latent heat of evaporation, long wave radiation, and heat conduction causes cooling at the sea surface (Ahrens, 2018). Even on a perfectly calm and clear day, the ocean's surface 1mm known as the surface skin, cools due to the latent heat of evaporation and long-wave radiation from the ocean surface to the surrounding atmosphere.

Evaporation rates are tied to the blowing wind over the sea surface. Offshore winds can have a significant impact on surface water temperatures. Because these winds are relatively warm in summer, they raise the ocean surface temperature and increase seawater evaporation. They absorb heat and cool the surface water near shore during the winter because they are much cooler than the ocean surface (Trujillo and Thurman, 2011).

Evaporation process is a key player in many physical operations, e.g. (1) As the rate of evaporation varies according to the area and location of the aquatic reservoir, and whether it is a sea or a lake, it will impact on the salinity of the basin. In arid regions where evaporation rates are extremely high, lakes can become hypersaline (Trujillo and Thurman, 2011). Because of their high evaporation rates, marginal seas like the Mediterranean and the Red Sea are crucial sites for the creation of dense water masses (Stewart, 2008; Wells, 2011). (2) Evaporation is a factor used to determine the type of basin whether positive, when the amount of water gained by precipitation and river runoff exceed that is lost due to the evaporation process, e.g. the Black Sea, or negative, when the reverse is true, e.g. the Mediterranean Sea. (3) The increase in the evaporation rates results in the increase of the water vapor in the atmosphere and hence in the relative humidity (Ahrens, 2018). (4) The increased evaporation rate raises the salinity at the surface, resulting in a halocline (Trujillo and Thurman, 2011).

On a local scale calculation of the evaporation rate from specified aquatic basins has been the core of a lot of research worldwide, of them:

The Arabian Gulf is marked by high SST associated with increases in evaporation rates. The computed mean monthly evaporation (2014-2015) showed a maximum in July (19.66 mm) and a minimum in January (0.915 mm) (Alfartusi, 2019). Due to its location in the arid climate zone, the Gulf evaporation (~1.84 m/yr) extensively exceeds water inputs (~0.30 m/yr) from precipitation and runoff (Ibrahim et al., 2020). The Red Sea is also featured by high evaporation and low precipitation (Al-Subhi, 2012). The average annual evaporation over the whole Red Sea Basin is ~2.0 m/yr (Morcos, 1970). This was also calculated by Sofianos et al. (2002) to be 2.06±0.22 m/yr. (Al-Subhi, 2012) showed that the rate differs between locations at Jeddah (2.03 m/yr), Gizan (1.28 m/yr) and Hodaidah (1.25 m/yr); referred to the period 1999-2007. Maiyza (1988) calculated this rate to be 1.28 m/yr at Hurghada City along the Egyptian Red Sea coast. In the Gulf of Aqaba, (Al-Taani et al., 2020) reported 4 m/yr; attributing this high rate to the subtropical desert location of the Gulf, with scare precipitation and absence of freshwater inputs. Over the period 1988-2003, (Kontogianni and Bartzokas, 2005) showed that the Ionian Sea has its lowest monthly evaporation in December (16 mm) and its maximum in July (153 mm). In the Adriatic Sea, the annual evaporation may reach 1.34 m/yr (Oddo and Guarnieri, 2011). Aegean Sea loses a minimum of 2 mm/day from its surface waters in spring and 5-7 mm/day in summer (Poulos et al., 1997). The maximum daily evaporation from the Mediterranean surface waters occurs in the Levantine (~4 mm/day) (Grodsky et al., 2019). (El-Geziry, 2021) calculated the minimum monthly evaporation from Alexandria Eastern Harbour (AEH) surface waters to be 6 cm in January and the maximum to be 14 cm in July and August over the period 2019-2020.

2.3.3.3. SST variations

The idea that ocean temperature fluctuations can influence atmospheric circulation at small scales (a few tens to hundreds of kilometers) has grown since the 1950s (Moulin, 2015). Variations in SST are primarily influenced by changes in the air–sea heat flux, as well as vertical mixing and horizontal heat advection (Skliris et al., 2012). Since the 1950s, the oceans have absorbed around 93% of excess heat attributable to anthropogenic global warming (Cheng et al., 2017), resulting in a strong increasing trend in global average ocean SST, as indicated by modern instrumental records (Deser et al., 2010). The influence of air sea-heat flux on the marine atmospheric boundary layer is responsible for much of the small-scale variability in the wind stress field. Previous research has found that natural climatic changes in mid-latitude regions are particularly susceptible to SST distributions on a diurnal, seasonal, inter-annual, and long-term basis (Kawamura et al., 2008). The diurnal fluctuations in the SST affect long-term occurrences and their variations, such as the Madden–Julian Oscillation (MJO) and ENSO (Shinoda, 2005; Shinoda et al., 2019). The diurnal variation in SST has also been studied to see whether there is any potential feedback on the atmosphere (Bernie et al., 2007; Kettle et al., 2009). SST at the ocean-atmosphere interface has a significant impact on society, for example through large ocean vortices and atmospheric circulation cells that affect weather and climate, weather systems, and severe storms and local events such as sea breezes and convective clouds. The SST variations have been the subject of many research in different regions worldwide, e.g. the whole Red Sea basin examined a SST increase over the period 1982-2016, with a rate of 0.029 °C/yr as concluded by Shaltout (2019). Rohrbeck et al. (2021) modelled the SST in the Red Sea basin over the period 1985-2015 and showed a positive trend for its SST anomalies with spatial variations of these anomalies over the whole basin. Pattiaratchi and Hetzel (2020) examined the SST variability along the eastern and western coasts of Australia (Fig. 2.18), over the period 1993-2016. Their results revealed strong seasonal SST on the two sides, with a general trend of increase attributed to the strong La Nina events in 1999 and 2011 when the mean monthly SST and SST anomalies were maxima.

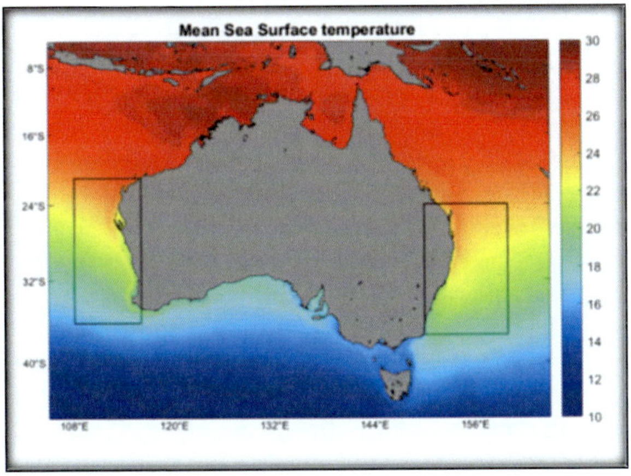

Figure 2.18. Mean SST along the eastern and western Australian coasts as depicted in (Pattiaratchi and Hetzel, 2020) over the period 1993-2016

A 138-year (1870–2007) monthly observed SST time series averaged over the ship track in the Indian Ocean running from the Gulf of Aden through the Malacca Strait showed a 1.4 °C warming throughout the whole period (Chowdary et al., 2014). Likewise, Roxy et al. (2020) showed that the Indian Ocean examined a general SST increase with a rate of 0.15 °C/decade over the period 1951-2015. With CO2 forcing at observed rates, the Coupled Model Inter-comparison Project (CMIP5) showed warming trends of 0.1–0.18 °C/decade from 1976 to 2005, with the highest warming trends over the northern Arabian Sea (Zhao and Zhang, 2016). However, despite the increasing trend in the Indian Ocean

SST, the net heat flux has a negative trend, contrary to expectations, which shows that local ocean dynamics and ocean-atmosphere interaction have a significant influence in the Indian Ocean's observed warming tendency (Rahul and Gnanaseelan, 2016). Li et al. (2020) examined the North Atlantic SST anomaly variations and the associated interannual variability in the North Atlantic Oscillation (NAO). At the interannual time scale, NAO contributes to generating the variabilities of the SST. The SST distribution in Aegean Sea has been investigated by Poulos et al. (1997) in relation to the prevailing heat fluxes and surface circulation. Their results revealed that SST in Aegean basin varies spatially and seasonally, reaching a maximum of 26 °C in summer (S) and a minimum of 8 °C in winter (N). The basin is characterized by a negative heat budget of -26 W/m^2 and a circulation mainly riven by wind conditions and seasonal thermohaline patterns.

Barale et al. (2004) used satellite images to research the SST in the Adriatic Sea over the period 1981-1999. Basic statistics for the north, central and southern sectors of the basin were obtained based on image analyses, revealing a SST increasing trend of 2.0 °C over 20 years, i.e. 0.1 °C/yr. From 1985 to 2008, Skliris et al. (2012) calculated an increasing SST trend in the Mediterranean Sea, with a rate of ~0.04 °C/yr. the same rate was reported for the Mediterranean by Pastor et al. (2018) for the period 1982 to 2016, and by Mohamed et al. (2019) for the period 1993–2017. (Shaltout and Omstedt, 2014) examined changes and anomalies in remotely sensed SST in the Mediterranean Sea from 1986 to 2005, and concluded a 0.24 °C/decade temperature increase, with a general shift in the Mediterranean SST from 13.8 °C to 22 °C. El-Geziry (2021) concluded a consistent warming trend for the SST in the southern Levantine basin (1948-2018), with a rate of 0.04 °C/yr, i.e., 0.4 °C/decade (Fig. 2.19).

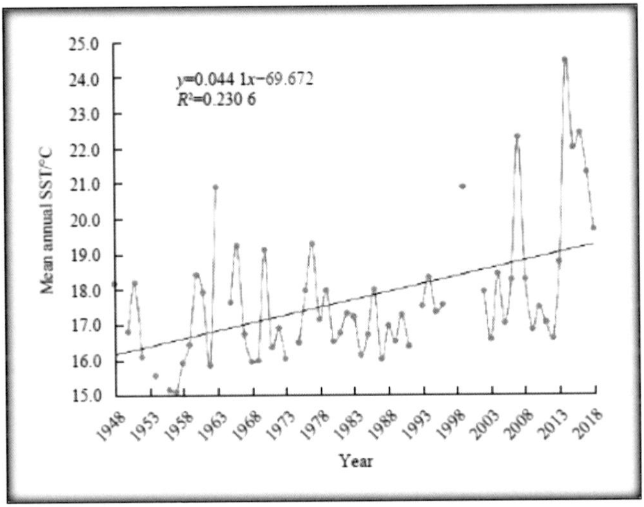

Figure 2.19. Trend of the mean annual SST in the southern Levantine basin over the period 1948-2018 (Source: El-Geziry, 2021)

2.3.3.4. Surges

The Surge is the local change in the sea level along a shore; attributed to atmospheric factors (atmospheric surges) and variations in hydrographic parameters (steric surges) (Eid, 1998). Surge heights are approximately hydrostatic in deep water, whereas surge production by wind stress and atmospheric pressure can be significant on shallow continental shelves (Idier et al., 2020). A surge height is derived by subtracting the astronomic tidal elevation from the recorded sea level and it typically has a duration of a few hours (Hicks et al., 2000). The total instantaneous elevation may substantially surpass the expected surge plus the astronomic tide when wind-generated waves ride on top of the surge. This is known as the storm surge. Storm surges have a potentially catastrophic impact, especially on low lying coasts. This destructive effect will be discussed in Chapter 3 of this book. The combined effects of SLR and surge are not yet included in regional hazard maps, although the consequences are expected to be significant (Tebaldi et al. 2012). Recent research has focused on the combined impact of storm intensification, surge heights, and the gradual increase in sea level rise (SLR). In the NW European shelf, atmospheric surges of tens of centimeters to one meter are common, e.g. Idier et al. (2012) and Breilh et al. (2014), and can approach 2–3 m under special stormy conditions, such as in the North Sea in 1953 (Wolf and Flather 2005). Nevertheless, atmospheric surge can reach about 10 m in cyclonic conditions (Nott et al., 2014). Investigation of the atmospheric surges has been the core of much research in different regions over the world, e.g. surges and atmospheric conditions along the Atlantic coast of France were examined by Pirazzoli (2000), who concluded less frequent occurrence of surges as a result of changing conditions in atmospheric pressure and wind regime over the period 1951-1997. The monthly variations in the atmospheric surge at Jeddah were said not to follow the hypothesis of decreasing 1 cm for every 1 mb increase in atmospheric pressure (Sultan et al., 1995; Sultan and Elghribi, 2003). Tsimplis et al. (2005) concluded that the inverse relationship to atmospheric pressure over the Mediterranean ranged from -0.58 to -0.68. More than 75% of surges in the Gulf of Lions are associated with a negative NAO (Ullmann and Moron, 2008). Orejarena-Rondón et al. (2018) examined the atmospheric surge in Colombia (Caribbean Sea) over the period 1991-2010 and concluded effective contribution of atmospheric pressure and wind on the observed sea level; leading to enhanced excess in the calculated MSL in the area of investigation. In Egypt, Moursy (1989) studied the surge behaviour in detail in Alexandria Western Harbour (AWH) over the period 1974-1983, and showed the effect of the variant atmospheric conditions on the changing behaviour of the calculated surge heights. Also, Eid (1989) showed that atmospheric pressure plays a key role in the observed variations in the surge heights off Alexandria coast. According to Saad et al. (2011), atmospheric pressure accounts for 16% of total sea-level variations in AWH, while wind accounts for 17%. Empirical equations to calculate the surge heights depending on the dominant atmospheric conditions at the two main harbours of Alexandria were derived by Eid (1998) for AWH and El-Geziry and Dabbous (2021) for AEH. The measured sea-level variations in Safaga Harbour on the Red Sea were more closely connected to air temperature and atmospheric pressure than to zonal or meridional wind speed (Fig. 2.20) on both hourly and monthly bases (El-Geziry et al., 2021).

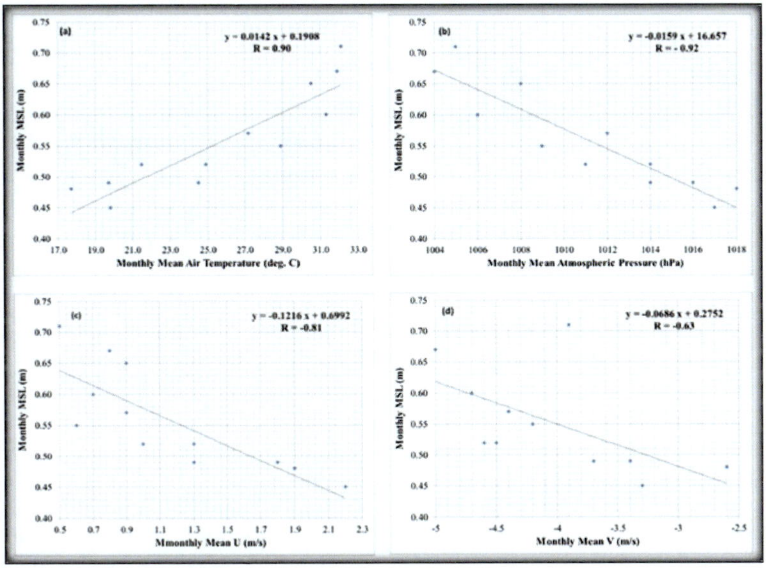

Figure 2.20. Correlation between monthly MSL and affecting atmospheric conditions in Safaga Harbour (2012-2014) (Source: El-Geziry et al., 2021)

2.4. Methods to measure/calculate air-sea interaction processes

Measuring air-sea interaction processes is still one of the most difficult tasks faced by meteorologists and oceanographers. This is mainly attributed to the difficulty of having continuous monitoring for the various factors controlling such processes (Rutgersson et al., 2020). As deduced from the previous sections, processes at the air-sea interface can be classified as turbulent or non-turbulent. The latter can be presented by radiative fluxes, e.g. solar insolation, evaporation or precipitation, whereas the former can be presented by wind-induced fluxes, e.g. surges or surface waves (Bourassa et al., 2010). Practically, factors controlling air-sea interaction processes can be measured either in coastal regions or over open seas (Csanady, 2004; Rutgersson et al., 2020). Micrometeorological in situ measurements of the air-sea can be taken from land-based towers (Rutgersson et al., 2020), air-sea interaction gliders (Fig. 2.21; Grare et al., 2021), drifting buoys (Graber et al., 2000; Chenbing et al., 2017) as in Figure (2.22), wave-riders (Reverdin et al., 2013), air-sea interaction profilers (Ward et al., 2014) and satellite imagery (Bourassa et al., 2010).

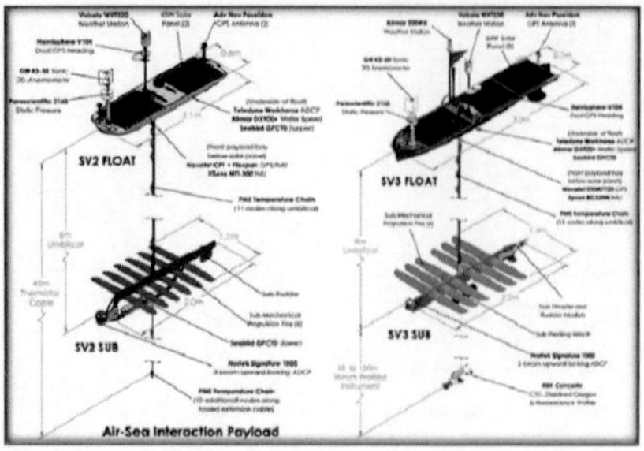

Figure 2.21. Views of air-sea interaction SV2 and SV3 Gliders (Source: Grare et al. 2021)

Figure 2.22. Air-Sea Interaction Spar (ASIS) buoy at the Sea (Source: Graber et al., 2000)

Nevertheless, one cannot only rely on the obtained data from such devices, which give information on amounts of solar radiation, rainfall, evaporation rates, SST, air temperature, atmospheric pressure, etc. Empirical equations have been the principal tool used to calculate the different fluxes at the air-sea interface, and in the recent years numerical models have come to play a vital role estimating the projected status of the different air-sea interaction processes. In this concern, equations used to calculate the surface heat fluxes will be given at a glance using selected approaches.

The net heat flux (Q_T) is calculated as the sum of four different fluxes (Wells, 2011) as shown in the following equation:

$$Q_T = Q_S - Q_L - Q_E - Q_H \text{ W/m}^2$$

Where, Q_S is the incoming solar radiation flux, Q_L is the longwave radiation flux, Q_E is the latent flux lost by evaporation and Q_H is the sensible heat flux. All are in W/m^2.

Q_S is calculated using the following equation (Reed, 1977):

$$Q_S = Q_{sc}\,(1 - \varepsilon C + 0.0019\beta)(1 - A)$$

Where, Q_{sc} is the daily mean values of incoming short wave radiation (W/m^2); ε is a constant taken as 0.62; C is the cloudiness cover, β is the noon solar altitude in degrees, and A is the albedo.

In latitudes from 20° S to 40° N, Q_{sc} is computed using a formula under clear sky conditions and a suggested albedo of 0.06 (Payne, 1972; Ramu et al., 2010; Hussein and El-Geziry, 2014). This formula is as follow:

$$Q_{sc} = A_o + A_1 \cos\theta + B_1 \sin\theta + A_2 \cos 2\theta + B_2 \sin 2\theta$$

Where A_s and B_s are latitude-dependent coefficients (Reed, 1977), and θ = (t-21)(360/365); t is the Julian day (day number)

Q_L is the amount of incoming radiation reflected at the ocean surface is determined by the sun's height and the state of the ocean surface as indicated by the ocean waves (Faizal and Ahmed, 2011). The Q_L term is affected by the ocean surface temperature, relative humidity, and cloud cover (Reed, 1977). It is calculated using the following equation (Faizal and Ahmed, 2011):

$$Q_L = 5.1 \times 10^{-8}\, SST^4\, [1 - (0.210 + 0.174 \times 10^{-0.055ea})(1 - 0.765\, C)]$$

Where, e_a is the actual water vapor pressure (mb).

On the basis of the vapor pressure at the sea surface, the actual vapor pressure in the air at a height of 10 m above the sea surface, and the wind speed at 10 m height, the rate of latent heat loss by evaporation (Q_E) from the sea surface is computed by using the following equation (Faizal and Ahmed, 2011):

$$Q_E = 1.4\,(e_s - e_a)W(2494 - 2.2SST) \times 10^{-3}$$

Where e_s is the saturated water vapor pressure, and W is the wind speed (m/s).

Lastly, the sensible heat flux (Q_H) depends on the difference between the SST and the overlying air temperature in addition to the wind speed measured at 10 m height above the MSL (Faizal and Ahmed, 2011), as shown in the following equation:

$$Q_H = -1.88\, W\, (SST - T_a)$$

Where T_a is the air temperature (°C)

CHAPTER THREE

EXTREME EVENTS

3.1. Definition of an extreme event

Despite various parties' efforts to get a consensus on a specific description of an 'extreme event' as the first step toward effective response, success in these initiatives remains elusive. The field of climatology is no exception, and both meteorologists and oceanographers are still missing a definite definition of an extreme weather/climate event. According to the International Panel for Climate Change (IPCC) in 2007:

> "An extreme weather event is rare at a particular place and time of year. Definitions of rare vary, but an extreme weather event would normally be as rare as or rarer than the tenth or 90th percentile of a probability density function estimated from observations. By definition, the characteristics of what is called extreme weather may vary from place to place".

In 2016, the World Meteorological Organization (WMO) gave a similar definition for an extreme weather/climate event:

> "The occurrence of a value of a weather variable above (or below) a threshold value near the upper (or lower) ends of the range of observed values of the variable".

Anyhow, despite the efforts of the international community, countries worldwide still require a specific description of extreme weather events as a first step toward proper reaction. Stakeholders should be aware that a severe weather event could occur at any time, with effects that could have been averted if we had been well prepared (Radovic and Iglesias, 2019). There is no doubt that the occurrence of such extremes has its severe impact on natural ecosystems, anthropogenic activities, human lives, and on the socioeconomic sector. According to Guha-Sapir et al. (2016), the number of meteorological disasters in 2016 reached a new high of 96 all over the world. The number of hydrological disasters (177) fell 6.9% below the decadal average (190.1), but the number of geophysical disasters (31) remained close to the annual average (31.6) of 2006-2015. On the other hand, the number of climatological disasters (38), which was 16.6% higher than the annual average (32.6), was the fourth highest since 2006.

There is evidence that anthropogenic activities, like as increases in GHG concentrations in the atmosphere, have caused some extremes to develop and shift, however, even if anthropogenic climate change did not exist, various natural weather and climate extremes would still occur (Field et al., 2012).

In this book, an extreme event is meant to describe abnormal occurrence of any processes induced by the air-sea interaction. Therefore, this may comprise, but not be limited to, droughts, flooding, heavy precipitation, hurricanes, tornadoes, storm surges, rogue waves, SLR, cold and heat waves, and so many other phenomena having various effects on the population and the environment, depending on their sensitivity to extreme events and their ability to recover from stress. Researchers are now confident that considering air–sea interaction processes improves the modeling of mean climate in the tropics, the portrayal of tropical precipitation extremes, monsoon prediction, and so many other climate phenomena (Dong et al., 2020).

Extreme events in air-sea interaction processes can be observed in so many phenomena, e.g. ENSO extremes, storm surges, extreme precipitation and droughts, tropical cyclones, tornadoes, rogue waves, hot and cold waves, flooding events, extreme SST and so many other events that exceed the normal occurrence. Hereafter, some of these extreme air-sea interaction events will be given at a glance.

3.2. Examples of extreme air-se interaction events

3.2.1. Extreme Precipitation and Droughts

An obvious example of extreme air-sea interaction event is global extreme precipitation and droughts affecting the different continents. Observations of precipitation have demonstrated increasing trends in the intensity of rain falling in the heaviest occurrences in many places, where "heaviest events" are commonly defined as those surpassing some high percentile of the distribution from some set of earlier years (Groisman et al., 2015). While changes in global mean precipitation are delimited by radiative processes, extreme precipitation events are both chiefly unrestricted by such global budget concerns and restricted by available moisture in the atmosphere, so that the amount of rain that falls in such events is further coupled to atmospheric water vapor content than is global (or even regional) mean precipitation (Sobel and Tippett, 2018). With increased global warming, heavy precipitation events are projected to intensify and become more frequent in most locations. Extreme daily precipitation occurrences are expected to intensify by roughly 7% on a worldwide scale for every 1.5 °C of global warming as shown in Figure (3.1) (Masson et al., 2021). Precipitation is expected to increase at high latitudes, in the equatorial Pacific, and in sections of the monsoon zones, but to decrease in the subtropics and in a few tropical locations (Masson et al., 2021). Extreme precipitation can put a strain on water treatment plants and sewage systems, as well as contribute to disease epidemics (Curriero et al., 2001). Heavy rains can increase the number of microbiological pollutants in runoff, posing a health risk (Parker et al., 2010). Intense storms can have a negative impact on agricultural production by severely damaging crops (Rosenzweig et al., 2002), have a negative impact on terrestrial ecosystems (Knapp et al., 2008), cause fatal landslides (Martelloni et al., 2012), increase the risk of infrastructure failure and damage (Nissen and Ulbrich, 2017), and worsen daily traffic conditions (Cools et al., 2010). The most significant impact of heavy precipitation, however, is the potential for severe floods, which could be exacerbated in urban areas and coastal regions affected by rising sea levels (Papalexiou and Montanari, 2019).

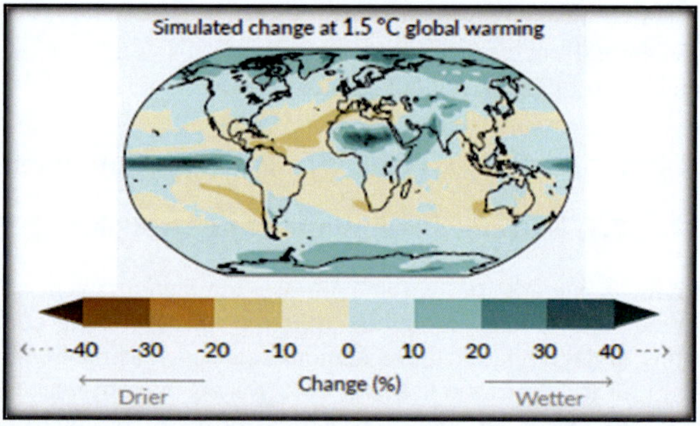

Figure 3.1. Projected annual precipitation change at 1.5 °C global warming relative to1850-1900 (Source: Masson et al., 2021)

Droughts are the polar opposite of severe precipitation, and are linked to climate in various ways. Droughts are classified into meteorological and hydrological draughts. Meteorological drought refers to long-term precipitation deficiencies inextricably linked to atmospheric circulation, and the changes projected because of global warming are, in essence, speculative (Sobel and Tippett, 2018). There is widespread agreement across models that meteorological droughts would become more common in particular specific regions as the climate warms, as well as some dynamical understanding of those changes, resulting in increased confidence in those areas (Masson et al., 2021). A prominent example is the Middle East, where increasing drought is strongly predicted by models and recent droughts have proven to be difficult to explain in terms of natural climate variability, implying an anthropogenic role (Kelley et al., 2015). The historical and paleoclimate records show very long and severe "mega-droughts," which are difficult to explain in terms of radiative forcing (Coats et al., 2016). Hydrologic drought, on the other hand, is described as a shortage of surface water reservoirs, including snow and soil moisture, and is controlled not only by precipitation but also by temperature, owing to temperature's influence on surface evaporation (Sobel and Tippett, 2018). Thus, the case for a human influence on hydrologic drought is compelling: while the lack of precipitation may be entirely natural in many circumstances, due to warming, evaporation, and snowpack loss, the surface water becomes even more depleted than it would be otherwise for the same precipitation deficit (Williams et al., 2015).

3.2.2. Rogue waves

Rogue waves (Fig. 3.2) are one of the oldest phenomena observed by mariners and sea-going people. This can be dated back to as early as 1826 when Captain Dumont d'Urville, a French scientist and naval officer in command reported confronting waves up to 30 m height (Kharif et al., 2009). In the maritime community, the term "rogue wave" has long been used to describe waves that are much higher than expected for the sea state, representing a terrifying and frequently life-threatening phenomenon (Dysthe et al., 2008).

Figure 3.2. The Italian Vessel 'Michelangelo' faced a 24 m height rogue wave in 12 April 1966 (Source: https://www.michelangelo-raffaello.com/en/michelangelo-12-april-1966/)

According to Trujillo and Thurman (2011), rogue waves are massive, solitary, sudden ocean waves that can reach incredible heights and frequently occur when normal ocean waves are not unusually large. When the wave height H (distance from trough to crest) is more than twice the significant wave height (H_s), the wave is referred to as a rogue wave (Dysthe et al., 2008). A significant wave height is defined as the average of the one-third observed largest heights (Kharif et al., 2009), and denoted as $H_{1/3}$ or H_s. Mariners frequently call rogue wave crests as "mountains" and their troughs as "holes", resulting in a dangerous rollercoaster ride. The relatively frequent occurrence of rogue wave events, as well as their spread across the world's oceans, leads to believe that the rogue wave phenomenon is related to the dynamics of typical waves on the sea surface, i.e., generated by the wind and affected by wind and atmospheric pressure while propagating (Kharif et al., 2009). Rogue waves are thought to be caused by an unusual occurrence of constructive wave interference, in which numerous waves overlap in phase to generate a massive wave (Spitz, 2021). Rogue waves can also be formed when strong ocean currents converge and magnify opposing swells. Such conditions exist off the southeast coast of Africa, where the Agulhas Current collides with large Antarctic storm waves, generating rogue waves that can crash onto a ship's bow, overwhelm its structural capacity, and sink even large, well-constructed ships (Trujillo and Thurman, 2011).

Rogue waves are unpredictable. They are only recognized when they occur. The chances of a truly rogue wave to occur in the open sea are only 1 in several billion. This is because that in the open sea, 1 wave in 23 will be more than twice the height of the ordinary wave, 1 in 1175 will be three times the height of the average wave, and 1 in 300,000 will be four times the height of the normal wave (Trujillo and Thurman, 2011). Rogue waves do, however, happen, and satellite measurements over a three-week period in the South Atlantic Ocean in the year 2004 proved that they happen more frequently than previously thought: More than ten distinct enormous waves of over 25 m in height were recorded by the study (Janssen and Alpers, 2006). Satellite radar research in 2007 confirmed that waves with crest to trough heights of 20 to 30 m occur significantly more frequently than previously anticipated (Smith, 2007). Rogue waves are now recognized to occur often in all of the world's oceans but even with satellite monitoring of ocean waves, predicting when and where rogue waves will occur remains difficult. Some rogue waves are so large that they endanger oil-drilling platforms and vessels at sea. 22 accidents due to collision with rogue waves were recorded over the period 1969-1994, as shown in Figure (3.3).

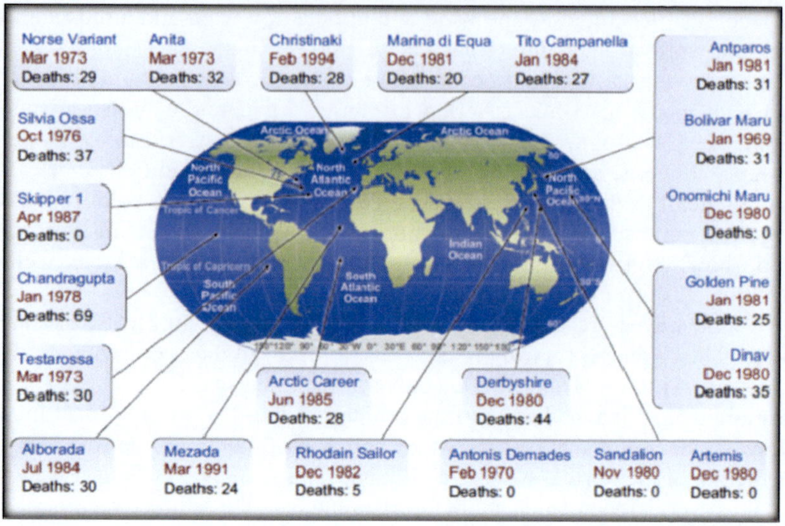

Figure 3.3. Statistics of rogue wave accidents recorded over the period 1969-1994 (Source: Dysthe et al., 2008)

3.2.3. Tropical cyclones

A tropical cyclone is an extreme air-sea event resulting from the interaction between a low-pressure system over surface tropical or sub-tropical waters, with organized convection and low-level winds that circulate either anti-clockwise (in the northern hemisphere) or clockwise (in the southern hemisphere) (in the southern hemisphere) (Oliver, 2005). A tropical cyclone is known by various names depending on its location and strength, such as hurricane, typhoon, tropical storm, cyclonic storm, tropical depression. While hurricane often refers to a tropical storm over the Atlantic and northeastern Pacific Oceans, typhoon refers to a severe tropical cyclone in the western North Pacific (Ahrens, 2018). According to the WMO (2017), around 80 to 90 named tropical cyclones form each year on average around the world, with more than half of them developing hurricane-force winds of 120 km/h or higher. Around the world, a tropical cyclone is considered to have formed when the mean surface winds exceed 65 km/h. At this point, it is assumed that a tropical cyclone has become self-sustaining and can continue to intensify without assistance from its surroundings. Many of these systems were unnoticed until they had impacted land or a ship had happened to come across them by accident before satellite imaging became available in the twentieth century.

Tropical cyclones usually form in the summer; however most tropical cyclone basins have seen them in virtually every month. The timing and frequency of tropical cyclone development are influenced by climate cycles such as ENSO. Tropical cyclones on either side of the Equator are typically formed in the ITCZ, where winds blow from the northeast or southeast. Air is heated over the warm tropical ocean and rises in discrete parcels within this broad area of low pressure, causing thunderous showers to form. This causes a flow of warm, moist, rapidly rising air, which begins to rotate cyclically as it interacts with the earth's rotation (Webb, 2021). A SST of around 27 °C and low vertical wind shear surrounding the system, atmospheric instability, and high humidity in the lower to middle levels of the troposphere, enough Coriolis force to develop a low-pressure center, and a pre-existing low-level focus or disturbance are all required for these thunderstorms to develop further (Berg, 2010). The following diagram (Fig. 3.4) displays the formation system of a tropical cyclone.

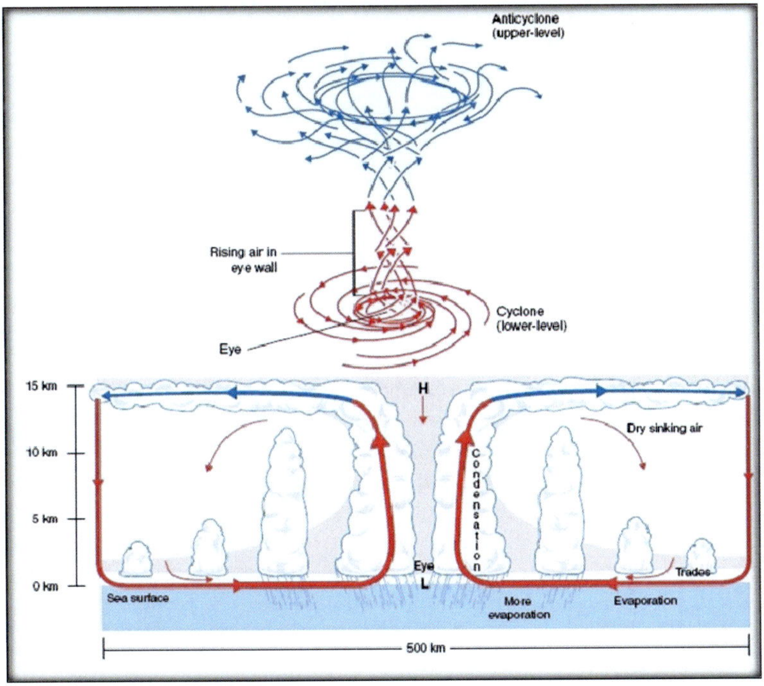

Figure 3.4. Wind flow and a cross-sectional view of a tropical storm (Source: https://cimss.ssec.wisc.edu/sage/meteorology/lesson5/)

Storm size is commonly measured using various metrics. The radius of maximum wind, the radius of outermost closed isobar (ROCI), and the radius of vanishing wind are the most commonly used metrics (Merrill and Gray, 1982). Another metric is the radius at which the cyclone's relative vorticity field drops to $1 \times 10^{-5}\ s^{-1}$ (Ahrens, 2018). Table (3.1) gives the size description of tropical cyclones based on the ROCI.

Table 3.1. The size description of tropical cyclones (Source: adapted from Merrill and Gray (1982))

ROCI (Diameter)	Type
< 2° latitude	Very small
2° - 3° latitude	Small
3° - 6° latitude	Medium/Average/Normal
6° - 8° latitude	Large
> 8° latitude	Very large

At the moment, tropical cyclones are officially named by one of twelve meteorological services and keep their names throughout their lives to facilitate communication between forecasters and the general public about forecasts, watches, and warnings (Schulz, 2017). The names are intended to reduce confusion if two storms form in the same basin at the same time. When storms reach sustained wind speeds of more than 61 km/h, names are usually assigned from predetermined lists based on the basin from which they originate (Cornish and Ives, 2009). Around the world seven warning centers exist in Northern and Southern Hemispheres, to issue advisories in their specified regions of responsibility; providing basic information and encompass a system's current, anticipated position, movement, and intensity. These centers are given in Table (3.2).

Tropical cyclones have severe impact on human lives and activities. Large waves, torrential rain, flooding, and high winds are caused by tropical cyclones at sea, interrupting international trade and, in some cases, causing shipwrecks. Strong winds on land can damage or destroy cars, buildings, bridges, and other external items, turning loose material into dangerous flying projectiles. Storm surge, or the rise in sea level caused by the cyclone, is usually the most dangerous effect of land falling tropical cyclones, accounting for 90 percent of tropical cyclone deaths in the past (Shultz et al., 2005). Tornadoes are produced by the broad rotation of a land falling tropical cyclone and the vertical wind shear at its periphery.

Tornadoes can also form because of eyewall meso-vortices that last until landfall (Oliver, 2005). From 30th September to 05th October 2021 Pakistan, Iran, Sultanate of Oman and United Arab Emirates were hit by tropical cyclone ‘Shaheen’ bringing heavy rain and winds of up to 150 km/h. On October 3rd, 2021, the Omani authorities reported 369 mm of rain at Al-Khaboura, which is located northwest of Muscat (the capital), and more than 200 mm in Muscat itself (Fig. 3.5). 13 people were announced dead: 11 in Oman and 2 in Iran (https://www.bbc.com/news/world-middle-east-58783992).

Table 3.2. Tropical cyclones warning centers (Source: WMO, 2017)

Tropical cyclone warning centers		
Basin	**Warning center**	**Area of responsibility**
Northern Hemisphere		
North Atlantic	U.S. National Hurricane Centre	N Equator, African Coast – 140° W
Eastern Pacific	U.S. Central Pacific Hurricane Centre	N Equator, 140–180° W
Western Pacific	Japan Meteorological Agency	Equator – 60° N, 180–100° E
North Indian Ocean	India Meteorological Department	Equator northwards, 100–40° E
Southern Hemisphere		
South-West Indian Ocean	Météo-France Reunion	Equator – 40° S, African Coast – 90° E
Australian region	Indonesian Meteorology, Climatology	Equator – 10° S, 90–141° E
	Geophysical Agency (BMKG)	
	Papua New Guinea National Weather Service	Equator – 10° S, 141–160° E
	Australian Bureau of Meteorology	10–40° S, 90–160° E
Southern Pacific	Fiji Meteorological Service	Equator – 25° S, 160°E – 120° W
	Meteorological Service of New Zealand	25–40° S, 160° E – 120° W

Figure 3.5. Tropical cyclone 'Shaheen' hitting Oman on 03rd October 2021 (Source: https://worldview.earthdata.nasa.gov/)

3.2.4. Strom surges

Storm surges are extreme changes in sea level caused by air-sea interaction through changes in atmospheric pressure and wind stress at the sea surface (Idier et al., 2020). Storm surges, which form during extreme events, can last anywhere from a few hours to three days and vary in time and space depending on whether the event makes landfall or passes, reaching heights of several meters along coastlines (Kim, 2019; Idier et al., 2020). Storm surges are classified as long waves with a time scale ranging from several hours to one day and a wavelength roughly equal to the width of the depression's center, which is typically between 150 and 800 km (Probstand and Franchello, 2012). The most dangerous events occur when an extreme surge coincides with a spring tide (Marcos et al., 2009). In this case, the combined water level is referred to as the storm tide level (Harper et al., 2001). This causes severe coastal flooding and puts the coastal environment at risk. Storm surges can be disastrous, especially on low-lying coastlines. Flooding caused by storm surges has the potential to harm not only coastal structures and human infrastructures but also human lives and the environment along the coast's wetlands, seagrass beds, deltas and shorelines. Storm surges can also have large impacts, as seen in the 1993 storm of the century, which affected much of the eastern United States (Thompson et al., 2013), Cyclone Xynthia in France in 2010 (Chadenas et al., 2014), and the North Sea flood in 2013 (Dangendorf et al., 2016). The historical evidence emphasizes the dangers of storm surges all over the world. Growing storminess increases the frequency and severity of extreme events in some parts of the world (Fortunato et al., 2016). Bertin et al. (2013), for example, demonstrated a significant increase in the significant wave height over the entire North Atlantic Ocean over the last century. On the other hand, vulnerability is growing as a result of the continuous migration of people and economic activity to the coast (Nicholls et al., 2008). This concern has resulted in significant legislation (the European directive on the assessment and management of flood risks, 2007/60/EC) and significant research funding in the 7th EU research framework and the Horizon 2020 EU research framework in Europe (Perini et al., 2015).

Historically, sea-level extremes have risen in synch with increases in mean sea level at coastal locations. Using this as groundwork, one can relate sea level extremes to mean sea level, allowing for the prediction of future extremes and return periods (Obeysekera and Park, 2013). Another approach is to calculate the probabilistic properties of the non-tidal residuals, NRT (component of storm surge and waves above the tidal variations). Nicholls et al. (2008) calculated and ranked the vulnerability of 136 coastal port cities with more than one million inhabitants in 2005 to coastal flooding caused by high wind and storm surge (Fig. 3.6). They identified broad-scale patterns of population and asset exposure in port cities, as well as how these patterns may change by the 2070s as economies and urban populations grow and the climate changes. The study also underlined the vulnerability of several of the rapidly developing cities to future SLR and flooding.

Recent research has focused on the combined impact of storm intensification, storm surge, and a gradual increase in the SLR. Physical factors have been evaluated using tide/storm surge models. Haigh et al. (2014), for example, presented a combined statistical-numerical modelling approach for estimating present-day extreme water level probabilities for Australia's entire coastline. The Worldwide Storm Surge Forecasting and Information System was developed by Buckman et al. (2015) as the first-of-its-kind operational forecasting system for storm surge prediction on a global scale, considering tidal and extra-tropical storm occurrences in real time. In regions where local storm surge prediction is currently lacking, the system gives real-time sea level and surge information. Muis et al. (2016)

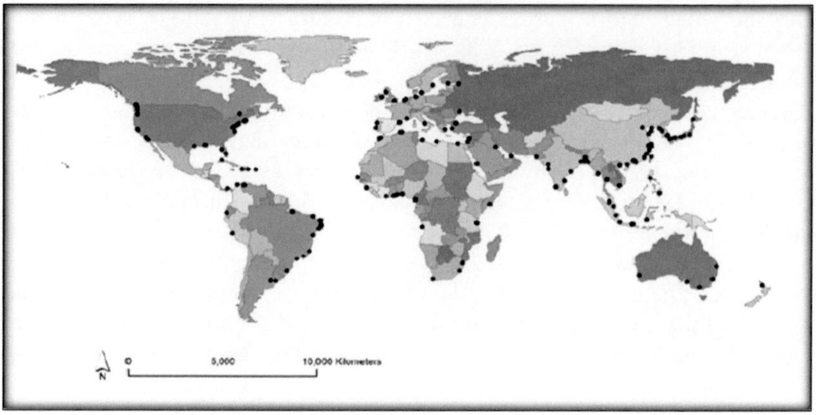

Figure 3.6. The location of the 136 port cities analyzed in Nicholls et al. (2008)

presented the first global reanalysis of storm surges, called Global Tide and Surge Reanalysis (GTSR), which was based on the hydrodynamic modelling and employed the Delft3D Flexible Mesh Suite with D-Flow. For the one-in-ten-year water level, they reported a root-mean-squared error (RMSE) of less than 0.45 m for 90% of the stations. Using a similar model, Vousdoukas et al. (2016) investigated the impact of climate change on extreme storm surge levels along of the European coastline by forcing the hydrodynamic model with wind and pressure fields from climate models. They reported RMSEs of less than 0.1 m for the majority of the Mediterranean, the Atlantic coast, and the Norwegian Sea during the validation period. Wahl and Chambers (2016) investigated the relationship between multidecadal extreme sea-level variation and large-scale climate variability along the US coastline using simple and multiple linear regression models. Cid et al. (2017) used multiple linear regressions to create a global storm surge database (spanning the years 1871 to 2010) that correlates mean sea-level pressure and gradients from the ERA-Interim reanalysis. Cid et al. (2018) reconstructed daily maximum storm surges for Southeast Asia between 1866 and 2012. They discovered a correlation of 0.7 or higher for 50% of the tide gauges in the region. Montblanc et al. (2019) demonstrated the implementation and validation of an unstructured hydrodynamic storm surge and tidal model-based pan-European storm surge forecasting system (EU-SSF). The average RMSE was reported to be less than 0.1 m, whereas extreme surge events (> 99th percentile) were simulated with an RMSE of less than 0.25 m. In addition to existing hydrodynamic numerical models, results of Tadesse et al.

(2020) emphasize the added usefulness of data-driven models in simulating storm surges at the global scale.

CHAPTER FOUR

MARINE HEATWAVES (MHWS)

While oceanic extremes have been researched for decades, most research has concentrated on a few important physical variables: (1) The majority of attention has been focused on exceptional sea levels, which have a lengthy history because of their connection to hydrology, and (2) Extreme wave heights and current speeds have piqued people's interest recently. Extreme events in water parameters (such as temperature, salinity, chlorophyll, and so on...), on the other hand, have gotten relatively little attention, despite the fact that they are likely to affect species and ecosystem distribution. Temperature extremes, in particular, have a significant impact on biological processes (Hobday et al., 2018). Extreme SST events detected around the world have been called marine heatwaves (MHWs). An MHW is defined qualitatively as a considerable period of abnormally warm water at a specific place, while quantitative definitions are based on ocean temperatures above a fixed (Frölicher et al. 2018), seasonally variable (Hobday et al. 2016), or cumulative (Eakin et al. 2010) threshold. MHWs are induced by a combination of local oceanic and atmospheric processes, such as air–sea heat flux and horizontal temperature advection, and can be influenced by large-scale climate variability, which can include remote sources via teleconnections (Hobday et al., 2016; Holbrook et al., 2019; Oliver et al., 2020).

The study of MHWs as discrete events is primarily motivated by their environmental and social consequences. In physical oceanography and climate research, the study of SST fluctuation has a long history, and understanding the consequences of warm ocean temperature extremes on marine organisms is critical (Oliver et al., 2020).

Over the last decade, MHWs have been reported in all major ocean basins, but only a few have been thoroughly described and researched (Frölicher and Laufkötter, 2018). The effect of MHWs can prolong and extend to impacts marine ecosystems on scales from hundreds to thousands of kilometers (Gupta et al., 2020). The frequency of MHW days has grown by more than 50% globally over the last century, owing mostly to long-term warming of the upper ocean (Oliver et al., 2018). Furthermore, even under scenarios of considerable GHG emissions abatement, future predictions show orders of magnitude increases in the number of MHW days at the end of the century (Frölicher et al., 2018). Numerous prominent MHWs have been linked to devastating changes in marine ecosystems around the world in recent years (Fig. 4.1). For example, MHW events in Australia in 2011, which lasted for more than 10 weeks, the NW Atlantic in 2012, the NE Pacific Ocean from 2013 to 2016 exhibited the 'Blob' MHW with a SST anomaly exceeding 6.0 °C, the 2015/16 MHW traversing the southeastern Indian Ocean to the Coral Sea, the Tasman Sea in 2015/2016 lasted more than 250 days, and the waters around tropical Australia in 2015/16 are examples (Frölicher et al., 2018; Holbrook et al., 2019).

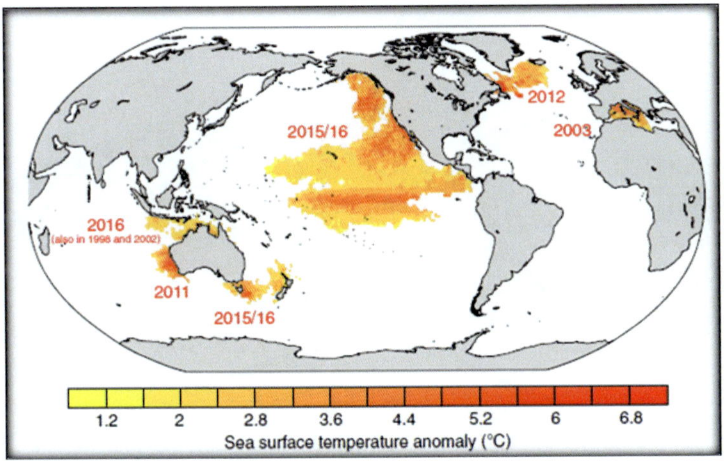

Figure 4.1. MHW events reported and analyzed in literature (Source: Frölicher et al., 2018)

The mechanism behind building-up MHWs, their persistence and decay is still poorly understood (Frölicher and Laufkötter, 2018). However, according to the available literature MHWs are frequently the result of a combination of regional heating and local weather factors that work together to promote fast heat accumulation (Fordyce et al., 2019). According to Gupta et al. (2020), three proximate causes can be set for initiating a MHW: (1) alterations in the ocean's heat transport, such as the amplification of boundary currents; (2) sustained large-scale atmospheric synoptic systems; and (3) coupled air-sea interaction processes. SST is the key player upon assessing any MHW. The heat sources and sinks inside the surface-mixed layer, which are reflected in SST changes, can be used to investigate the physical processes that cause MHWs (Oliver et al., 2020). The development, evolution, and decay of MHW have been described using a mixed-layer temperature budget (Kataoka et al. 2017). Temperature fluctuations in the surface mixed layer are linked to physical processes such as horizontal temperature transport and air–sea heat fluxes using this method (Oliver et al., 2020).

Temperature anomalies can be quantified using sea surface temperature percentiles and account for a wide range of temperature regimes in different marine environments. Marine heatwaves can be classified into four categories (Fig. 4.2) based on their intensity when temperatures exceed the 90th percentile over a 5-day period (Hobday et al., 2018). Figure (4.3) depicts the statistics of the global MHWs from 1982 to 2019.

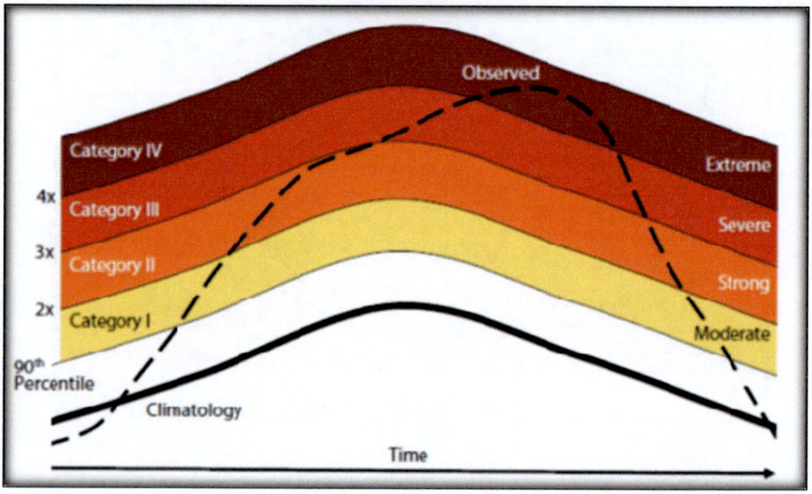

Figure 4.2. Categories of MHWs showing the observed SST (dashed), Climatological reference (bold) and 90th percentile SST (thin) (Source: Hobday et al., 2018)

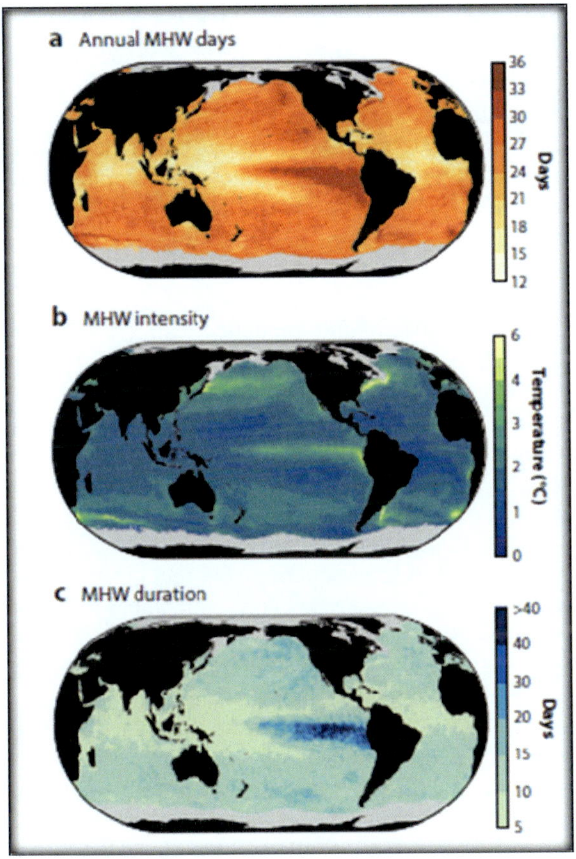

Figure 4.3. Global MHWs (a) annual number days of occurrence, (b) SST anomaly (°C) and (c) duration (Source: Oliver et al., 2020)

Mathematically, the rate of change in SST in the mixed layer affecting the occurrence and duration of a MHW is obtained by the following equation (Song and Yu, 2017):

$$\frac{\partial T}{\partial t} = -u.\nabla T + \overline{\nabla.(k_h \nabla T)} - \frac{1}{h}k_z\frac{\partial T}{\partial z} - \left(\frac{T - T_h}{h}\right)\left(\frac{\partial h}{\partial t} + u_h.\nabla h + w_h\right) + \left(\frac{Q_S + Q_L + Q_E + Q_H}{\rho C_p h}\right)$$

Where, T is the temperature in the mixed layer, t is the time, u is the velocity vector in (x, y) plan; w is the vertical velocity, h is the mixed layer depth, k_h and k_z are the horizontal and vertical diffusivity coefficients, respectively, C_p is the seawater specific heat capacity, ρ is the seawater density and the sum of Q is the air-sea heat fluxes (refer to section 2.4). The left-hand side term denotes the temperature tendency. The right-hand side denotes the horizontal advection (first term), the horizontal mixing (second term), vertical mixing (third term), the entrainment advection and induction processes (fourth term) and the air-sea heat flux (fifth term)

MHWs can be simulated and modelled using the Coupled Model Intercomparison Project-5 (CMIP5) using the IPCC Representative Concentration Pathways (RCP) scenarios as examined in Oliver et al. (2019). According to the projections, MHWs will become more severe and longer throughout the 21st century.

MHWs have dramatic effects on the marine ecosystem. In the not-too-distant future, MHWs will emerge as powerful agents of disruption to marine ecosystems; as they have already done on land, during terrestrial heatwaves with enormous consequences on the mortality of birds and mammals and significant repercussions on human health. According to the most comprehensive synthesis of MHW impacts to date (Smale et al., 2019), MHWs, regardless of their climatological attributes, all had a negative effect across taxa (including seaweeds, corals, birds, and mammals) and processes (such as growth, reproduction, and survival). Marine populations are most vulnerable to MHWs when they live near their maximum thermal limit, because temperature anomalies here are more likely to exceed organismal physiological thresholds, increasing mortality rates (Smale et al., 2019). Dramatic losses of marginal populations at equatorward range edges are anticipated with increased MHW activity, as has been documented for seaweeds (Thomsen et al., 2019), seagrasses (Thomson et al., 2015), and invertebrates (Garrabou et al., 2009). Anthropogenic activities like fisheries, aquaculture production and coastal structures, as well as human health will also be severely influenced by the intensified MHW occurrence (Oliver et al., 2019).

CHAPTER FIVE

EXTREME AIR-SEA INTERACTION EVENTS IN THE MEDITERRANEAN SEA

5.1. On the morphometry of the Mediterranean Sea

The Mediterranean Sea basin is located between 30° 00' and 45° 00' N and 06° 00' and 35° 00' E. The topography of the Mediterranean Sea and the names of its principal sub-basins are depicted in Figure (5.1). The Mediterranean basin is a vast semi-closed basin connected to the Atlantic Ocean by the Strait of Gibraltar, which measures 13 km in length and has a sill depth of 300 m at its westernmost point (Drakopoulos and Lascaratos, 1999). By way of the Dardanelles/Marmara Sea/Bosphorus system, the Mediterranean Sea is connected to the Black Sea. Europe (N), Africa (S), and Asia (E) encircle the Mediterranean Sea. The Mediterranean Sea shelf is narrow, with mountain chains dipping sharply into the sea on its northern flanks, resulting in a limited coastal zone and small drainage basin (Millot and Taupier-Letage, 2005). The Mediterranean Sea has a surface area of 2.5 x 10^6 km^2 (Pinet, 2009), accounting for 0.82% of the world's ocean surface area (Bianchi and Morri, 2000). The Mediterranean shores span 46 x 10^3 km, with an average depth of 1.5 km. In the Mediterranean Sea, the Ionian Sea has the deepest depth of 5.1 km (Pinet, 2009). The 400-meter-deep Sicilian-Tunisian sill splits the Mediterranean basin into two significant sub-basins: western and eastern basins, and serves as a geographical and hydrological boundary (Zervakis et al., 2003). The western basin is 0.85 x 10^6 km^2 in size, whereas the eastern basin is 1.65 x 10^6 km^2 in size (UNEP, 1996).

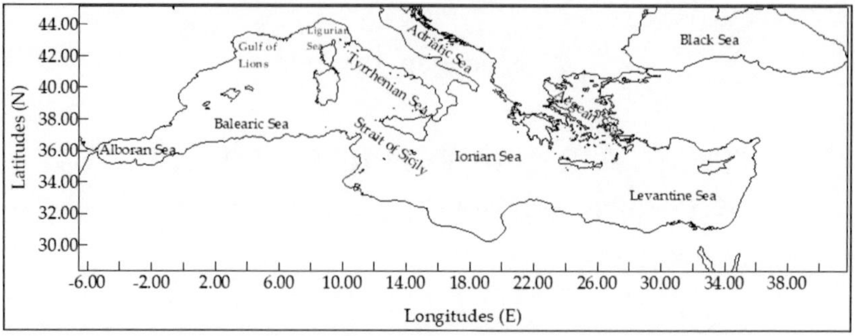

Figure 5.1. Geography of the Mediterranean Sea and nomenclature of its sub-basins

Figure (5.2) depicts the topographical features and bathymetry of the Mediterranean Sea basin with its two wings. The Alboran Sea, in the furthest westerly zone between Spain and Morocco, and the Balearic Sea, off the Algerian beaches, stretching north to the Gulf of Lions, make up the western Mediterranean basin. The Ligurian and Tyrrhenian Seas are located to the north and east of this western basin, respectively. The continental shelves in the western basin are narrow, except the Gulf of Lions (Millot and Taupier-Letage, 2005). The Tyrrhenian Sea, located between the Italian Peninsula and the islands of Sardinia and Corsica, is the western basin's easternmost and deepest component, measuring about 3500 m deep, and connecting the remainder of the basin to the south in a large gap between Sardinia and Sicily (Zavatarelli and Mellor, 1995). The eastern Mediterranean basin is more complex than the western basin. There are four sub-sea basins to be found there: the Ionian, Levantine, Adriatic, and Aegean Seas. The Ionian Sea is a body of water lying between Italy and Greece to the north and Libya and Tunisia to the south, with a maximum depth of 5000 m south of Greece (Millot and Taupier-Letage, 2005). This eastern basin's eastern zone is occupied by the Levantine Sea. The Levantine Sea has depths ranging from 2500–3000 m in the basin's center to 4500 m in a depression southeast of Rhodes Island (Horvat et al., 2003). The Adriatic is the Mediterranean Sea's most continental basin, bounded by two mountain chains (the Apennines and the Balkans) and extending latitudinally (Russo and Artegiani, 1996). The Strait of Otranto, with a width of around 75 km and a depth of about 800 m, connects it to the Ionian (Zavatarelli and Mellor, 1995). Several passageways between Greece, Turkey, Crete, and Rhodes connect Aegean Sea to the Levantine Sea. Its maximum depth is roughly 1500 m, and its beaches and topography are quite uneven.

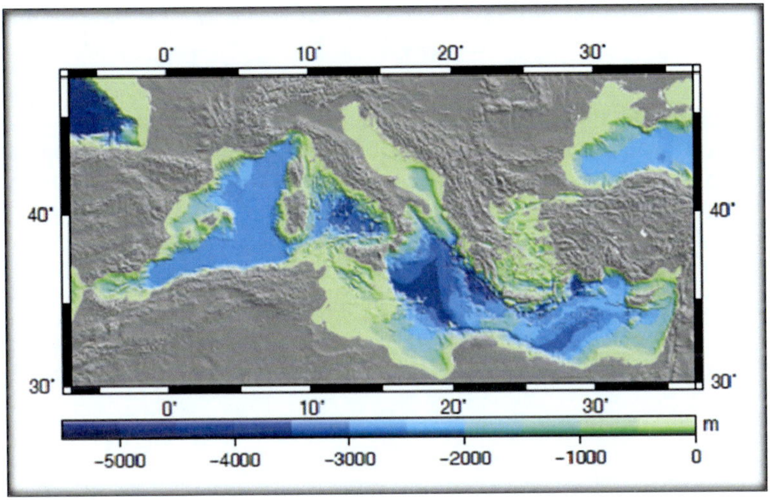

Figure 5.2. Bathymetry and topographic features of the Mediterranean Sea (Source: Arabelos et al., 2010)

5.2. Examples of extreme air-sea interaction events in the Mediterranean Sea

5.2.1. Storm surges

In the Mediterranean coastal regions, storm surges pose a major threat. Positive surge events are added to tide levels in the Mediterranean coasts, increasing the risk of coastal flooding due to excessive water levels (Cid et al., 2016). There are several low pressure systems over the Mediterranean Sea, some of which may acquire a dynamical evolution akin to tropical cyclones (medicanes) (Cavicchia et al. 2014). Storms like these cause severe disasters on the densely populated coastal areas that surround the Mediterranean Sea. As a result, it is critical to try predicting future storm surge impacts on the shore in general coastal planning and flood risk assessments. The severity of future coastal floods is determined by the SLR, land motion, and marine storminess (Conte and Lionello, 2014). The NAO has a strong correlation with surges in the western and central Mediterranean (Marcos et al., 2009), whereas the relationship with surges in the eastern Mediterranean is equally strong but weaker due to the distance from the atmospheric (pressure) action centers (Makris et al., 2015).

Krestinitis et al. (2011) used a 90-m horizontal resolution digital elevation model to evaluate the SLR caused by storm surge events in the NE Mediterranean coastal zone between 2000 and 2004. They were successful in identifying the low-lying places in the eastern basin that could be affected by such a rise (Fig. 5.3). The greatest sea-level height in the Adriatic Sea, particularly along the SE Turkish and Cyprus shores, topped 1 m in 2004. While the bulk of Greek beaches exhibited SLR ranges of 0–0.5 m, the Levantine-Aegean passage examined maximums greater than 0.5 m.

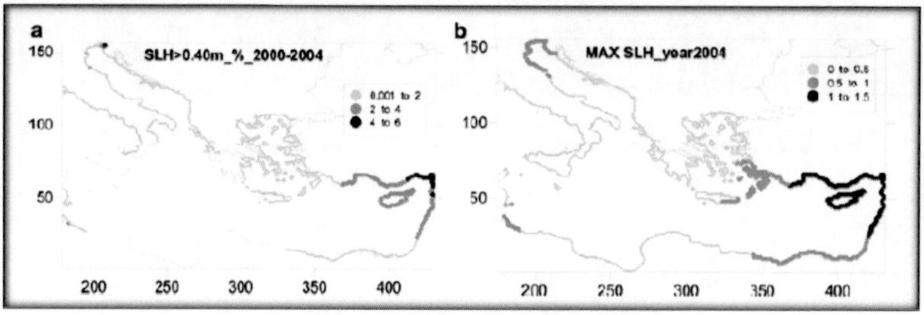

Figure 5.3. (a) % of occurrence of sea-level heights (SLH)>0.40 m in the eastern Mediterranean Basin and (b) the maximum SLH (m) in 2004 (Source: Krestinitis et al., 2011)

Conte and Lionello (2014) conducted research on future changes in storm surge patterns over the Mediterranean using the IPCC's A1B climate scenario. They used a shallow water hydrodynamic model. Their findings revealed that the largest surges occur in the eastern Mediterranean, specifically the Gulf of Gabes and the north Adriatic Sea. Makris et al. (2015) investigated the trends of sea-level extremes caused by atmospheric conditions in the Greek seas over a 150-year period (1951-2100) in a future climate scenario with rapidly growing atmospheric greenhouse gas concentrations. Their findings revealed that the bulk of extreme occurrences occurs largely in the winter and, to a lesser extent, in the spring. However, they discovered that due to increased cyclogenesis, summer extremes are increasing, particularly in southern areas. Similarly, autumn extremes may be more common over the N-Adriatic region, owing to higher southerly wind occurrence frequencies in the fall. Cid et al. (2016) examined how extreme storm surges in southern Europe have changed in terms of frequency, strength, and length. They detected more than five events per year over the Mediterranean's western and eastern basins, including the Levantine basin. The length of these surges is significant in the eastern Mediterranean. Positive rates of 1.5 h/yr are found in particular along the southern Italian and Greek coasts. Similar results are obtained in the Levantine basin, the Alboran Sea, and along the Turkish and Egyptian coasts. Along the Egyptian Mediterranean coast, storm surges were studied since the 1970s. Moursy (1976) investigated the storm surges along the coast of Alexandria city over the period 1965-1969. She found a strong correlation with the NW wind blowing with speed of more than 10 m/s. Hamed (1983) evaluated the number of stormy days in Alexandria during a 20-year period and found that the stormiest days occurred in the winter and early spring, with a significantly lower number in the autumn and no storms during the summer. Storm surges caused by winter storms over the period 1971-1984 in the eastern Mediterranean, near the Alexandria coast, were categorized into five types by Hamed and El-Gindy (1988) according to synoptic circumstances. The height ranged between a lowest of 12 cm (Type A) to more than 43 cm (Type E). The multiple regression analysis approach was used by El-Gindy et al. (1992) and Eid (1998) to develop prediction equations of storm surge heights in AWH. A powerful storm that battered Egypt's Mediterranean coast on 11th December 2010 (Fig. 5.4), with peak winds of more than 16.6 m/s. With a surge of almost 1 m, this storm produced coastal flooding in Alexandria. This value is far higher than the average storm-surge readings observed off the Nile Delta coast (40 to 50 cm). Fortunately, the storm only caused minor flooding in the lowlands of Abu-Qir Bay, east of Alexandria, and minor damage to certain coastal structures (Said et al., 2012). Hendy et al. (2021) calculated the storm surge height at Alexandria and concluded that a storm height of 76.2 cm can be reached once in 100 years and of 81.3 cm once in 200 years. Ibrahim (2021) concluded that storm surges with elevations higher than 65 cm accounted for 20% of the calculated surge heights at Mersa Matruh in 2004.

Figure 5.4. Impact of storm surge over Alexandria in December 2010
(Source: Tonbol et al., 2018)

5.2.2. Meteotsunamis

Meteotsunamis are atmospherically induced tsunamis, an extreme air-sea interaction process, which have destructive impact on coastal communities and are known since ancient times (Pattiaratchi and Wijeratne 2015; Vilibic et al. 2016; Vilibic et al., 2021). Meteotsunamis are featured by sea level fluctuations of several meters with periods varying from two minutes to two hours, and can arise without warning still during calm weather (Monserrat et al., 2006; Vilibic et al., 2021). Meteotsunamis have had an influence on a huge number of coastal settlements across the globe, except in Antarctica where there is no evidence of occurrence (Williams et al., 2019; Vilibic et al., 2021), e.g. the East US Coast, from Florida to Maine (Wertman et al. 2014), the South African coast (Okal et al. 2014), the Great Lakes (Bechle et al. 2016), the English Channel and UK coast (Williams et al. 2019). They also have severe impacts on human lives and cause serious injuries, e.g. the 2014 Odessa meteotsunami (NW Black Sea) when six people were injured (Sepick et al., 2018) and the 2017 meteotsunami in the Arabian Gulf which resulted in one death and five missing persons (Salaree et al., 2017). Despite this, the destruction and property damage generated by meteotsunamis is limited to a few hotspots (harbours, bays, and beaches) and is far less than that caused by destructive seismic tsunamis due to the reduced wave heights and geographical range of meteotsunamis compared to the greatest tsunamis. Furthermore, human casualties from meteotsunamis are uncommon and only occur in the most extreme cases, but human casualties from big seismic tsunamis, particularly those causing ocean-wide impacts, can be significant (Gusiakov et al. 2019). The physics of meteotsunami waves and their atmospheric sources have been increasingly researched in recent decades, thanks to increased availability of high-frequency (1 minute) sea level and meteorological data, as well as the development of numerical tools accompanied by an increase in high-performance computing resources (Vilibic et al., 2016).

Pressure and wind stress from moving atmospheric weather systems trigger meteotsunamis (Monserrat et al. 2006). A resonance mechanism is responsible for the meteotsunami's intense amplification (Rabinovich, 2009). External resonance in the open sea between propagating air pressure disturbances and sea-level oscillations, followed by internal resonance in the harbor between sea level oscillations and fundamental mode, is known as the double resonance effect (Monserrat et al., 2006). According to Williams et al. (2019), resonance between the meteotsunami and atmospheric forcing (external resonance) can provide up to an order of magnitude amplification throughout two potential external resonances, namely: the Greenspan resonance and the Proudman resonance. The Greenspan resonance occurs when the atmospheric forcing speed along the coastline equals the speed of a coastally trapped edge wave, whereas the Proudman resonance occurs when the atmospheric forcing speed equals the speed of a shallow-water wave (Monserrat et al., 2006; Williams et al., 2019).

Meteotsunamis have traditionally been studied in the Mediterranean region (Fig. 5.5) far more than elsewhere in the world (Vilibic et al., 2021). This is mainly due to the Mediterranean Sea micro-tidal nature, with tidal amplitudes (a few tens of centimeters except in the Gulf of Gabes and the northern Adriatic Sea) of an order of magnitude less than meteotsunami amplitudes. The primary focus of these studies has been on air sources, energy transfer to the sea, and meteotsunami wave propagation and amplification. This comprises, but not limited to, the work of Tintore et al. (1988) who investigated the meteotsunami events in the Balearic Islands in the 1980s and 1990s; the meteotsunami event in 1978 and the 2000s in the Adriatic Sea were studied by Vucetic et al (2009) and Vilibic and Sepic (2009), respectively.

Although famous with the occurrence of meteotsunamis, the Mediterranean basil is still missing a meteotsunami detection/warning system, which is mainly attributed to the cost of such systems to get in force. Factually, only two products have been thoroughly tested and published to date: the Balearic RIssaga Forecasting System (BRIFS) depending on records of 1 minute sea level from 6 tide gauges and atmospheric records from 20 air pressure sensor sites (Renault et al. 2011) and the Adriatic Sea and Coast (AdriSC) system depending on a pilot microbarograph network comprising of three air pressure sensors with a sample rate of one minute, utilized to detect strong air pressure disturbances in real time. The air pressure rate of change is measured every 5 minutes in this system, and when a meteotsunamigenic disturbance is detected, the strength, period, speed, and direction of propagation are automatically estimated. (Denamiel et al. 2019 a,b).

Figure 5.5. Number of meteotsunami events in the Mediterranean Sea (Source: Vilibic et al., 2021)

5.2.3. Rogue waves

As mentioned in section 3.2.2, rogue or freak waves represent an extreme air-sea interaction process face by mariners with wave heights exceeding 2 Hs. Rogue wave incidents have been observed and recorded in the Mediterranean basin. A 10.32 m high rogue wave was measured by a Datawell Directional Waverider buoy moored at 85 m in the Black Sea, according to Pelinovsky et al. (2004). Sulisz et al. (2016) looked at data from the Datawell Directional Waverider buoy, which is moored at 20.5 m depth in the Baltic Sea's southern region. They discovered several waves that met rogue wave statistical thresholds between 1996 and 2002, with the majority reaching up to 6.2 m in height and the highest detected extreme waves reaching up to 12 m. Though, over the years, a modest number of rogue wave sightings and reports have been reported in the Mediterranean Sea. The upper decks of a cruise ship named "Voyager" were damaged by 14 m high waves in February 2005 near Sardinia, according to the crew (Knobler et al., 2021). In March 2010, another cruise ship, the "Louis Majesty," experienced an extremely violent sea 38 km off the Spanish coast when two persons were killed and several more were injured in the incident (Cavaleri et al., 2012). At the time of the event, the Hs was roughly 4 m (5 m as recorded by the nearest buoy, 25 km distant), although eyewitnesses reported three waves exceeding 8 m crashing into the ship (Cavaleri et al., 2012; Innocenti et al., 2021). Cavaleri et al. (2021) reported on an unusually high wave captured by a web camera on an oceanographic tower 15 km off the coast of Venice, Italy, on 15th December 2009. The estimated crest amplitude was 5.1–6.4 m, which met the statistical rogue wave criterion. The only observational research on rogue waves near the coast of Israel to date was reported (Glozman and Sladkevich, 2019) during a large storm on 19 January 2018 by analyzing waves data provided by a nearshore deployed buoy at Ashdod, indicating a possible occurrence of a rogue wave with a height of more than 7 m (Fig. 5.6), assuming the rogue wave only occurs in extremely high sea events.

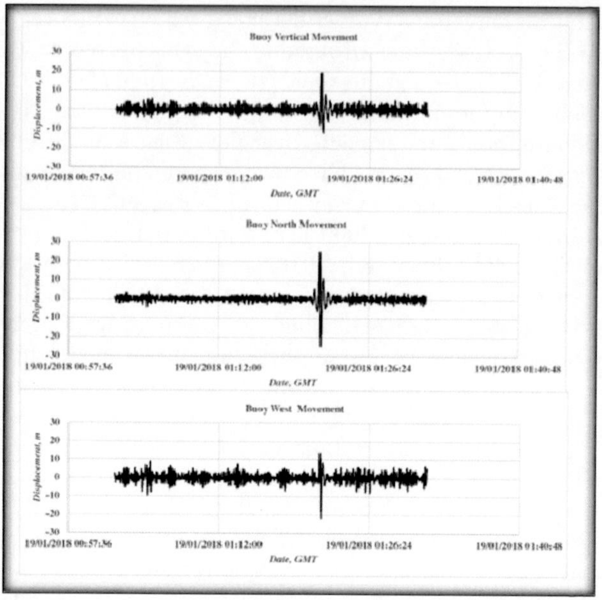

Figure 5.6. Buoy vertical displacement at Ashdod on 19.01.2018
(Source: Glozman and Sladkevich, 2019)

5.3. Case Study: Marine Heatwaves in the Mediterranean Sea

The Mediterranean Sea is no exception in the examination of MHWs. Depending on the GHG emission scenario, the annual mean Mediterranean SST by the end of the 21st century is anticipated to rise from +1.5 °C to +3.0 °C above current levels (Adloff et al. 2015). This considerable increase in SST is likely to hasten the occurrence of future MHWs, in line with forecasts of a 200 to 500% increase in GHG-induced heat stress across the region (Diffenbaugh et al. 2007).

To date, our understanding of historical MHWs in the Mediterranean Sea is based primarily on mass benthic species deaths connected to thermal anomalies, which have been increasing since the early 1990s (Darmaraki et al., 2019). The most notable example that has been extensively investigation is the MHWs occurred in 2003 and hit the NW Mediterranean basin. Due to considerable increases in air temperature and a reduction in wind stress and air-sea exchanges, SST anomalies of 2.0–3.0 °C above climatological mean persisted for nearly a month. What really happened is that in the summer of 2003, Europe was hit by a record-breaking heat wave (particularly in France). According to Sparnocchia et al. (2006), the Ligurian Sea's August surface temperatures were as high as 28-28.5 °C in 2003 (2-3 °C higher than prior summers) and uniform in both the central and coastal zones. According to Olita et al. (2007), anomalous June SSTs of up to 2.5 °C had a substantial impact on SSTs in the Tyrrhenian Sea (absolute SSTs over 29.5 °C). The mean SST anomalies observed over the period 18th July – 03rd September 2003 is shown in Figure (5.7). The NW Mediterranean was the mostly affected area by these MHWs. According to the daily SST anomaly variations, this extreme event was said to be ran at the upper depth of the sea surface over four phases, with phase 3 being the largest extended for one month from 03rd August 2003 to 04th September 2003, as depicted in Figure (5.8). As seen, this is much shallower than the normal mixed layer depth.

Because MHWs are usually caused by high air temperatures, their effects are usually limited to the top of the water column (Galli et al., 2017). Therefore, warming was limited to the upper layers (above 20 m depth), with the lower layers being abnormally cold. This was discovered to be caused by abnormal heat flow, which was produced in part by high air temperatures (the atmospheric heatwave) and in part by low wind speeds. This maritime heat wave was most likely responsible for the widespread mortality that occurred in nearby rocky reef communities at the time (Garrabou et al., 2009).

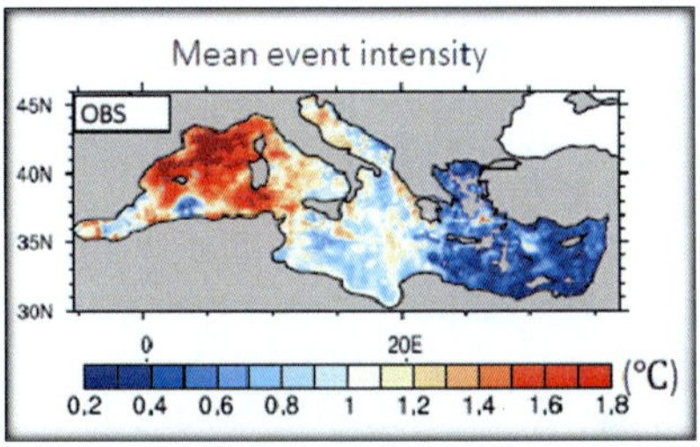

Figure 5.7. Observed SST anomalies over the Mediterranean during 48 days from 18.07 to 03.09.2003 (Source: Darmaraki et al., 2019)

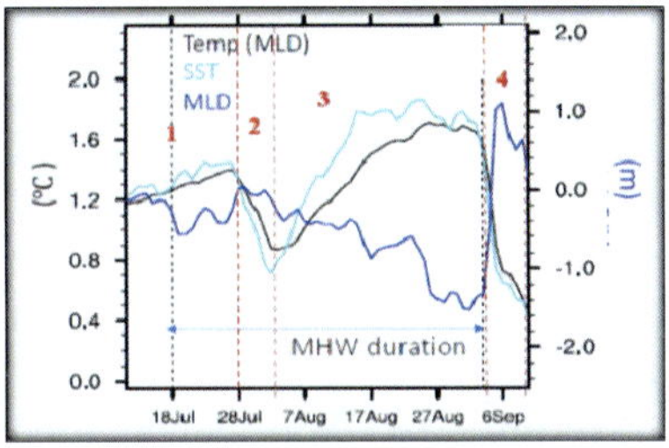

Figure 5.8. The four phases of the MHW event over the Mediterranean in 2003 (Source: Darmaraki et al., 2019)

The European heatwave in the summer of 2003 coincided with anomalous SSTs in the Atlantic Ocean, which were constructed in a roughly zonal banded pattern of positive and negative anomalies, according to Black et al. (2004). This occurred at the same time as record SSTs in the

Mediterranean (temperatures exceeding 30 °C). Sutton and Hodson (2005) related Atlantic Ocean basin-scale fluctuations (i.e., the Atlantic Multidecadal Oscillation) to midsummer climate in North America and Western Europe. These MHWs impacted on the Mediterranean ecosystem and caused mass mortality eruptions in a diverse hard substrate macroinvertebrates including anthozoans, sponges, bryozoans, ascidians, and bivalves, and resulted in a loss of seagrass meadows, and rapid alterations in maritime community structure (Galli et al. 2017; Darmaraki et al., 2019).

Unfortunately, there has been little research on the MHWs published in the eastern Mediterranean basin, and the subject is still being contested (Pastor et al., 2020). Recently, a study investigated the MHWs in the eastern Mediterranean basin over 39 years (1982-2020) using SST records (Ibrahim et al., 2021). The authors identified and examined the most severe MHW episodes that occurred in the Basin throughout the study period, as well as their link to meteorological conditions. From 1 January 1982 through 31 December 2020, daily SST data were acquired from the National Oceanic and Atmospheric Administration's Optimum Interpolation Sea Surface Temperature (NOAA-OISST), on a 0.25° grid scale. Atmospheric conditions of wind regime, air temperature, atmospheric pressure, in addition to MSL were obtained from the European Centre for Medium-Range Weather Forecasts (ECMWF) ERA5 at the same spatial resolution. The mean SST seasonal cycle was calculated from the mean monthly value over the calendar 1982-2020 at each grid point. Trends of annual and interannual SST variations were produced using an Empirical Orthogonal Function (EOF). Linear trends were obtained by using the least squared method and the statistical significance of the obtained results at 95% confidence was obtained using the Mann-Kendall test. Results revealed that over the eastern Mediterranean basin, there was a lot of spatial variability in the SST linear trend, ranging from 0.1 to 0.5 °C/decade. The average warming rate in the basin was 0.33±0.04 °C/decade. The Aegean and Levantine Basins had the highest values, while the Ionian and Tyrrhenian Seas had the lowest values. In all locations, there was a clear decadal increase in MHW frequency, as shown in Figure (5.9 a, b). Most of the Aegean and Levantine Seas, south of Crete and in front of Libya's eastern coast, had a frequency of 1–2.5 counts with a maximum significant ($p > 0.05$) value of > 2 events. On the other hand, results showed that MHW frequency did not change significantly ($p > 0.05$) SE of Crete, in front of Mersa Matruh city, or N of the Ionian Sea region. For the duration of occurrence, the MHW mean duration rose significantly ($p > 0.05$) with values > 17 days in front of the Egyptian

Mediterranean coast in the SE Levantine Sea and off the Israeli coast, with a tendency of >10 days per decade. In SE Crete, off Mersa Matruh city, and the north Ionian Sea region, there are no significant results ($p > 0.05$) for MHW duration (Fig. 5.9 c, d). SSTs were significantly related to the different atmospheric conditions affecting the air-sea interaction processes over the Basin. Two extreme MHWs were detected over the period of investigation: 24.04.2013-08.05.2013 (15 days) at 40.875° N 24.375° E, and from 28.08.2020 to 31.12.2020 (126 days) at 34.375° N 34.625° E, with an SST increase of +5.0 °C and +3.41 °C, respectively, over the normal mean.

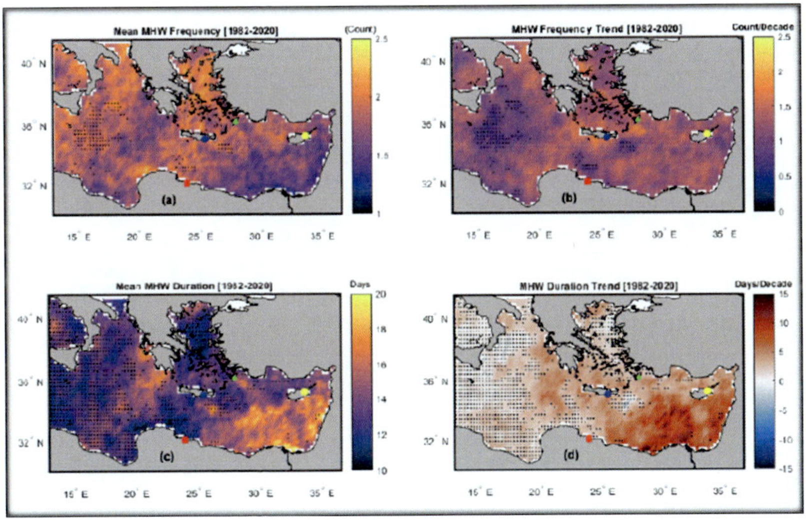

Figure 5.9. MHWs over the eastern Mediterranean Basin (a/b) frequency of occurrence and (c/d) duration of occurrence over the period 1982-2020 (Source: Ibrahim et al., 2021)

To the best of the author's knowledge, taking the above mentioned results on the MHWs affecting the eastern Mediterranean basin in consideration, the future research on this extreme air-sea interaction event should pay more attention to the anthropogenic elements that contribute to MHW episodes being more severe and recurrent, and to the probable MHW connects to several meteorological and atmospheric indices impacting the Mediterranean basin such as the North Atlantic, East Atlantic, and Mediterranean Oscillations. Moreover, investigation of the impacts of such extremes on the Mediterranean ecosystems is crucial and essential. This is a must in order to integrate between the various driving factors leading to experience the event, and to learn more about the detection, characterization, effect assessment, and prediction of MHWs; aiming to build a robust conclusion about the MHWs behavior and how can we minimize their negative impacts.

CHAPTER SIX

EXTREME AIR-SEA INTERACTION EVENTS IN THE RED SEA

6.1. On the morphometry of the Red Sea

The Red Sea (Fig. 6.1) is an adjacent sea to the Indian Ocean, between Africa and Asia. It is connected to the Ocean at its southern extremity through the Bab El-Mandab Strait and the Gulf of Aden. To the north, the Sinai Peninsula divides the Red Sea water body into the Gulf of Aqaba (E), and the Gulf of Suez leading to the Suez Canal (W). The basin can be, therefore, divided into three main water bodies: The Red Sea Proper, the Gulf of Suez and the Gulf of Aqaba. The Sea is underlain by the Red Sea Rift, which is part of the Great Rift Valley. Seven countries border the entire Red Sea basin, six of which border the Red Sea Proper with the Kingdom of Saudi Arabia having the longest shoreline along its eastern side and Egypt along its north western coast (Morcos, 1970). Jordan, the seventh country, borders the eastern side of the Gulf of Aqaba. The Red Sea extends longitudinally between 12° 30' N and 30° 00' N, with a total length of 1932 km, 280 km average width (Abdel Wahab, 2010), and approximate 4.58 x 10^5 km^2 surface are (Head, 1987). The Red Sea is divided into three distinct depth zones: shallow shelves with a depth of less than 50 m, deep shelves with a depth of 500 to 1000 m, and the central axis with a depth of 1000 to 2900 m (Madah et al., 2015). A detailed bathymetric map is depicted in Figure (6.2).

The Gulf of Aqaba is the NE arm of the Red Sea, connected to the Red Sea Proper through the Straits of Tiran. The Gulf is described as a deep, narrow, and partially-isolated coastal water basin (Al-Taani et al., 2020). Even with the harsh environment, the Gulf of Aqaba sustains unique aquatic habitats and biodiversity, as well as one of the world's most diverse coral groups (Al-Rousan et al., 2007), despite the fact that it is one of the Middle East most anthropogenically influenced coasts (Al-Taani et al., 2013). This includes coastal constructions, ports activities, dredging and reclamation processes, oil pollution, industrial wastes and others

(Batayneh et al., 2014). The Gulf of Suez, on the other hand, is the NW arm of the Red Sea, connected to the Red Sea Proper through the Strait of Jubal. The importance of the Gulf stems from being the main gateway connecting the Red Sea to the Mediterranean Sea through the Suez Canal at its northern extremity. The Gulf, covering an area of approximately 10000 km^2, is a shallow basin of about 300 km long and almost 65 m depth over its main axis (Gad et al., 2019). At the Gulf's entrance, there are plenty of islands and coral reef barriers, with the bottom sloping to depths of over 500 m (Morcos, 1970). The Gulf of Suez's coastal region is one of Egypt's most extensively industrialized areas. The Gulf of Suez is exposed to several sources of contamination, which can be classed as sewage, organic materials, heavy metals, oils, nutrients, sediment mobilization, and litter (Mahmoud et al., 2020).

The Red Sea proper, the main water body of the Red Sea basin, is characterized by its bathymetric variations, which influence its marine habitat. In the shallow water, there are a few-kilometer-wide bordering reefs and barrier reefs, while coral reefs dot the deeper water slopes randomly (Rasul et al., 2015). The continental slope has an uneven profile, with a succession of steps leading down to a depth of approximately 500 m, and the bathymetric depressions known as "Deeps", generated by the spreading of the sea, are found in the 15% of the Red Sea that constitutes the narrow axial trough (Rasul et al., 2015). Mountain ranges border the Red Sea Proper on both sides. This orography has a significant impact on the local dominant wind regimes, which are distinguished by the winter and summer seasons (Langodan et al., 2014). Underwater earthquakes occur beneath the Red Sea proper at varying depths south of latitude 27.8° (El-Isa, 2015). The Red Sea Proper hosts many anthropogenic activities, e.g. petroleum industry, touristic landmarks, diving sites, ports and harbours, etc. These are mainly concentrated in its northern part along its both eastern and western coastlines.

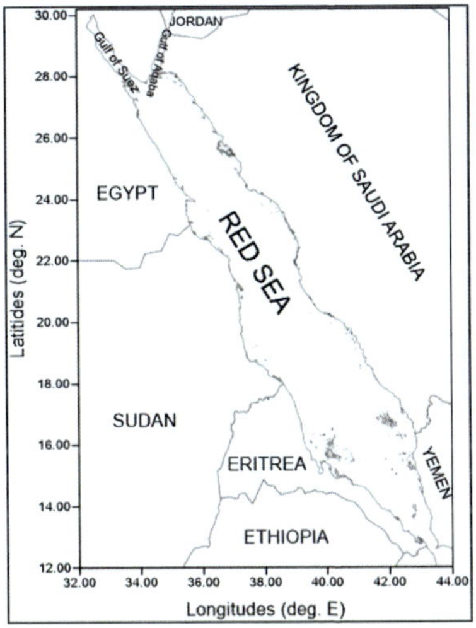

Figure 6.1. Map of the Red Sea basin, the two Gulfs of Suez and Aqaba and the surrounding countries

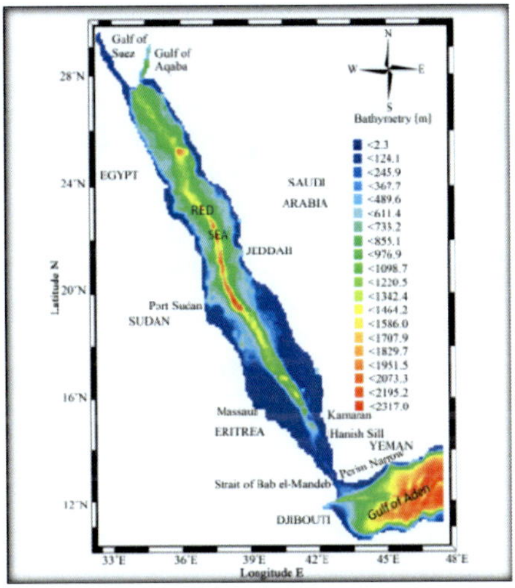

Figure 6.2. Bathymetry and topographic features of the Red Sea (Source: Madah et al., 2015)

6.2. Examples of extreme air-sea interaction events in the Red Sea

6.2.1. Extreme surface waves

The surface wave climate has been found as one of the most sensitive markers of local wind climate in semi-enclosed basins (Shanas et al., 2017). The Red Sea's wind climate is influenced by reversing monsoons (Jiang et al., 2009), and climatic fluctuations are represented in the primary monsoons: SW and NE (Ralston et al., 2013). However, due to the structure of the basin, its declined major axis prevents monsoon waves from the Indian Ocean from entering the Red Sea, and hence it is considered that swells from the Indian Ocean have little impact on the Red Sea (Langodan et al., 2014). Furthermore, in the center Red Sea, wind convergence is critical for controlling surface wave behaviour (Aboobacker et al., 2016). The Red Sea surface waves are moderate, with a maximum H_s, generally, of less than 4.0 m and with 2.0-2.5 m being more common (Ralston et al., 2013; Langodan et al., 2017) as shown in Figure (6.3).

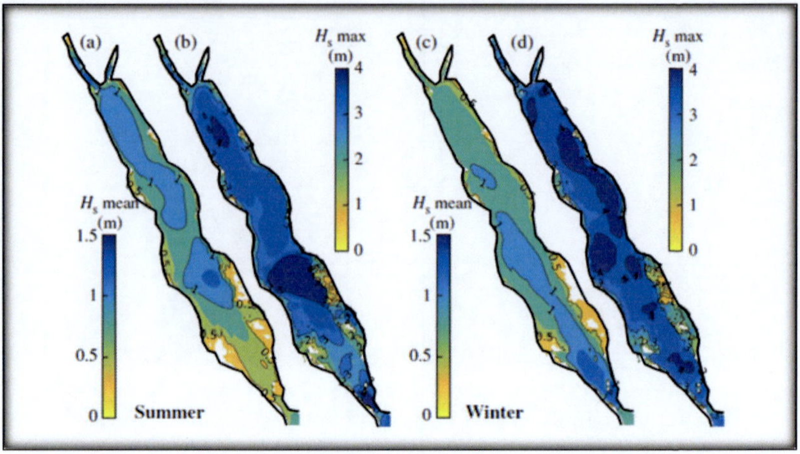

Figure 6.3. Mean and maximum Hs in the Red Sea in summer (a/b) and winter (c/d) seasons (Source: Langodan et al., 2017)

The typical H_s varies spatially between 0.6 and 1.2 m, with average peak periods of 4-6 s (Shanas et al., 2017). It is logical to assume that surface waves in the Red Sea are mostly related with dominating wind regimes; in summer, they primarily flow south, while in winter, they converge toward the basin's center. The multiple transversal wind jets caused by mountains running down the coastline disrupt this seemingly orderly structure, potentially resulting in temporarily increased wave heights (Langodan et al., 2014). Factually and surprisingly, only one wave buoy takes place in the whole Red Sea basin (Ralston et al., 2103; Langodan et al., 2014; Shanas et al., 2017), belonging to the King Abdullah University of Science and Technology (KAUST), and deployed at 22° 10' N, 38° 30' E (Fig. 6.4).

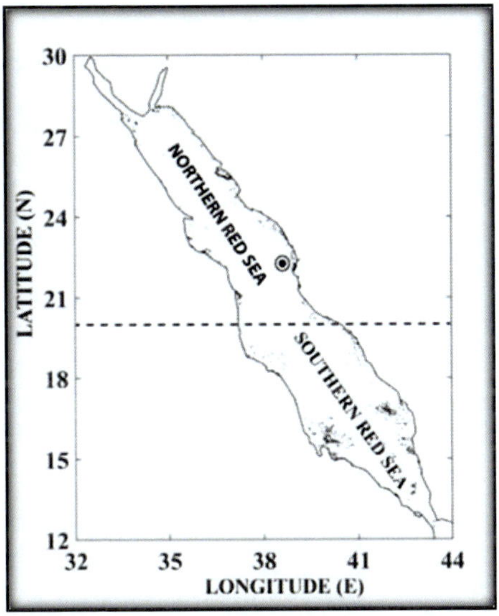

Figure 6.4. Location of the deployed wave buoy by the KAUST in the Red Sea (Adapted from Shanas et al., 2017)

Along the Red Sea basin, except for the shallow sections bordered by corals, the extreme (99th percentile) and near extreme (90th percentile) Hs follow the wind speed pattern, as the Red Sea region is a wind dominated sea region (Shanas et al., 2017). The extreme Hs is more common in winter with strong positive trends of winds dominating the season, especially in the central zone of the Red Sea (Langodan et al., 2017; Shanas et al., 2017). According to the investigation of Shanas et al. (2017), the greatest extreme $H_s \geq 4.0$ m heights (99th percentile) are clustered around 23° N, 19° N, and 15° N latitudes, whereas the 90th percentile of $\geq$ 2.5 m height are found in deep Red Sea waters. According to Langodan et al. (2017), the strongest seasonal winds flowing from the African side, associated with the Indian Summer Monsoon (ISM), cause the largest waves in the Red Sea's central region. These extremes were tightly correlated with the wind regime over the Red Sea basin with wind speed higher than 8 m/s (Shanas et al., 2017).

Dabbous (2020) discussed the seasonal variations in the H_s extremes for the 95th and 99th percentiles of occurrence along the Egyptian Red Sea coast over the period 2009-2015. She divided the coast into northern (Hurghada), middle (Safaga and Qusseir) and southern (Mersa Alam) sectors, and concluded that the southern region of the Egyptian Red Sea coast exhibited the maximum 99th percentile extreme H_s in three seasons: 2.1 m (winter), 1.8 m (spring), and 1.6 m (summer), whereas the northern region had the maximum extreme Hs in autumn (1.8 m). The same pattern extends for the maximum 95th percentile extreme H_s, with the southern region extreme H_s of 1.6 m in winter, 1.35 m in spring and 1.2 m in summer, and 1.35 m in autumn in the northern region. The middle sector has exhibited the lowest extreme H_s among the three sectors for both the 99th and 95th percentiles. Again, all these extremes are associated with the NNW wind impacting on the Egyptian Red Sea coast, with speed exceeding 10 m/s.

Extreme surface wave heights in the northern and southern Red Sea are expected to occur more frequently under anticipated climate circumstances from 2070 to 2099, compared to current conditions, whereas a lower frequency is anticipated in the middle Red Sea (Shanas et al., 2017).

6.2.2. Extreme SST

According to Al-Subhi and Al-Aqsum (2008) excess evaporation over precipitation, air temperature, wind regime and associated current patterns, land and topography influences, and daily and seasonal weather changes all influence SST fluctuations in the Red Sea. From south to north and from east to west, the temperature drops and the Arabian side of the Sea has greater temperatures to the southeast and east, whereas the African side has lower temperatures (Al-Barakati et al., 2002).

Warming's effects are most likely to be felt in semi-enclosed seas, which have a higher rate of warming than average and where organisms' ability to adapt to warming by shifting their biogeographical range poleward is limited by the presence of continental masses, making most semi-enclosed seas climatic sink areas for marine organisms (Chaidez et al., 2017). Therefore, identification of Red Sea (semi-enclosed basin) locations where SST may surpass high limits is critical because SST has a huge environmental and ecological impact on ecosystems of vulnerable animal species like as corals, for example (Lewandowska et al., 2014; Chaidez et al., 2017; Hazra and Huser, 2021). The Red Sea is considered one of the extremely warm sea basins that are rapidly warming (Raitsos et al., 2013). The average annual temperature of the Red Sea increased by 0.74 °C between 1982 and 2006, comparable to the global average of 0.85 °C (Belkin, 2009). In 1994, a strong warming event resulted in a 0.7 °C increase in the mean annual Red Sea SST (Raitsos et al., 2011).

Based on daily SST records, Chaidez et al. (2017) characterized the variability in SST extremes (T_{max}) across the Red Sea over the period

1982-2015. Their results revealed that the Red Sea has a latitudinal gradient of increasing T_{max} from N to S, with the southernmost Bab El-Mandeb Strait having the highest T_{max} (33 °C). T_{max} occurs in the northern Red Sea throughout July, while T_{max} in the southern Red Sea occurs between late July and mid–August. The calculated rate of T_{max} increase over the study period was 0.17±0.07 °C/decade. Hazra and Huser (2021) identified regions of extreme SST along the Red Sea using satellite imageries at a 0.05° x 0.05° resolution. Their results revealed that extremely SST may occur during the next century, potentially inflicting significant ecological damage to endemic species (such as particular types of corals) as well as economic ramifications for adjacent countries. Nevertheless, within this century, large areas of the southern Red Sea, including key coral reefs, are likely to undergo combined severe SST episodes (Fig. 6.5), which could have serious effects in terms of coral bleaching and mortality.

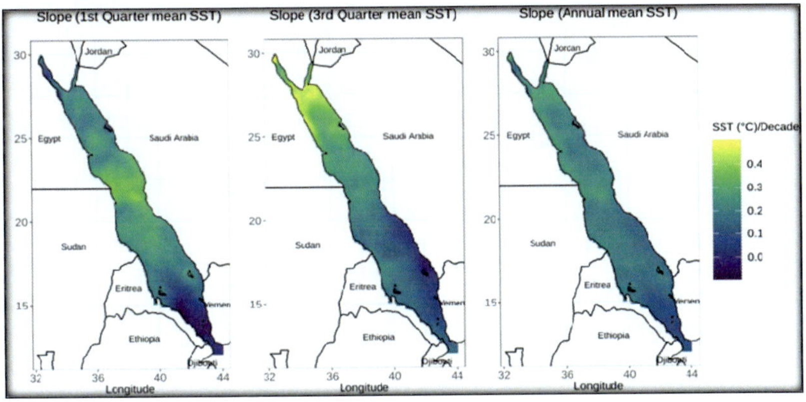

Figure 6.5. Extreme SST projection for the next 10 (left), 20 (middle) and 50 (rights) years for September-October period based on 2020 records (Source: Hazra and Huser, 2021)

6.2.3. Extreme precipitation rates

All over the Red Sea Basin, the mean amount of rainfall and the number of rainy days are relatively low and scarce (Benestad et al., 2012). Even though, flash floods can occur in the Middle East as a result of extreme precipitation events, which can have major socioeconomic consequences. Flash floods, for example, killed about 600 people in Egypt in November 1994 and more than 120 people in Jeddah (Saudi Arabia) in

November 2009 (De Vries et al., 2013). Extreme precipitation events over the Red Sea usually occur during the wet season, which extends from November to January (Alharbi, 2018). The Red Sea Trough (RST), a low-level monsoon trough at 850 hPa, is one of the key drivers of extreme precipitation over the Red Sea (De Vries et al., 2013; Dasari et al., 2017) by transforming moistures from the southern Red Sea to the Red Sea Convergence Zone (RSCZ) in the middle of the Basin (Dasari et al., 2017). The interannual variability in both RST and RSCZ is influence by a number of factors (Krichak et al., 2012; De Vries et al., 2013): (1) Locations of the RST and two high pressure zones over NE Africa and the Arabian Peninsula; (2) the intensity of the ENSO and the IM; (3) moisture transport and convergence and (4) large scale forcing and tropospheric instability resulting in mesoscale convective systems and (local) extreme precipitation. The RST is said to occur more frequently in autumn than other seasons (Krichak et al., 2012). Nearly 80% of the moisture from the Red Sea is transmitted throughout the Arabian Peninsula by upper layer transportation (above 850 hPa), which is closely tied to the Arabian high pressure (Zolina et al., 2017).

Research on extreme precipitation along the Red Sea was more intense on the Saudi sites, e.g. Jeddah event in 2009 (Alkhalaf and Basset, 2013; Haggag and El-Badry, 2013; De Vries et al., 2016); Jeddah event in 2010 (Yesubabu et al., 2015); Yanbu event in 2009 (Haggag and El-Badry, 2013); Gizan event in 2010 (Tekeli and Fouli, 2016). On the western bank of the Red Sea such studies are scarce whether in Egypt or Sudan. A recent study by Baldi et al. (2020) examined the extreme precipitation over Sinia Peninsula and considered two locations on the Gulf of Suez (Ras Sedr) and the Gulf of Aqaba (Newabaa) as two out of four locations exhibited extreme precipitation over the period 1997-2014. Results revealed that Ras Sedr examined 9 events, while Newabaa examined 6 events.

6.3. Case Study: Marine Heatwaves in the Red Sea

The Red Sea is expected to be vulnerable to SST warming trends during the current century, which could have a negative impact on marine entities and ecosystems, e.g., the deterioration of the average coral colony areas (Hoegh-Guldberg et al., 2014). Raitsos et al. (2011) confirmed that the Red Sea SST rose rapidly in the mid-1990s. For the warmest (August) and coolest (February) months, the Red Sea SST showed a significant trend of +0.5 and +0.38 °C/decade, respectively, over the period 1950-2007 (Hoegh-Guldberg et al., 2014).

Over the period 1982–2015, MHWs were observed more frequently in the northern half of the Red Sea (Chaidez et al., 2017). According to Shaltout (2019), the Red Sea exhibits MHWs in ~3% of the year (~10 days/yr) with most of which in the Gulf of Aqaba. He detected 16 MHW events along the entire basin of the Red Sea over the period 1982-2016, the minimum of which lasted for four days off Port Sudan (14.10.2002 to 17.10.2002) and the maximum for 18 days in the Gulf of Aqaba (29.07.2012 to 15.08.2012).

Recently, a study investigated the MHWs in the Red Sea basin over 38 years (1982-2019) using high resolution (0.05° x 0.05°) daily SST records from the Copernicus Marine Environment Monitoring Service (CMEMS) website from 1 January 1982 to 31 December 2019 (Mohamed et al., 2021). The mean SST seasonal cycle was calculated from the mean monthly value over the calendar 1982-2020 at each grid point. Trends of annual and interannual SST variations were produced using an EOF. Linear trends were obtained by using the least squared method and the statistical significance of the obtained results at 95% confidence was obtained using the Mann-Kendall test. The authors used the prolonged anomalously warm-water events with at least 5 days duration in a location and a 90th percentile threshold; to specify events of MHWs. Results revealed a trend of increase in the SST over the 38 years, with a rate of 0.32±0.03 °C/decade. Spatially, except for the southernmost part of the Red Sea near Bab El-Mandab and the Gulf of Suez, the MHW maximum intensity showed values of >1.8 °C throughout most of the Red Sea. The MHW greatest significant trend values (p >0.05; >2.5 °C) were discovered off the Egyptian coast near to the Gulf of Suez in the northern Red Sea and in the middle region of the Basin. On a temporal scale, from 1982 to 1994, the annual mean MHW frequency was less than 1, while from 1994 to 2019, the frequency increased significantly to reach its peak of 7 events in 2018. No MHW event was detected in 1984, while in 2019 the longest MHW event was detected. This lasted for 234 days from 10.05.2019 to 31.12.2019 at 29.625° N; 32.375° E, with 2.94 °C above the mean (Fig. 6.6). The authors recommended further investigation on the causes of MHWs and their influence on the marine ecosystems in the Red Sea.

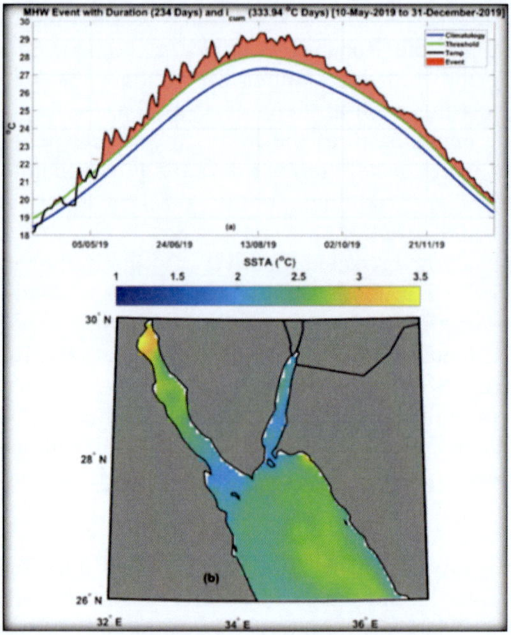

Figure 6.6. MHW event in the northern Red Sea in 2019 extended over 234 days (Source: Mohamed et al., 2021)

Coral bleaching is the worst impact by a MHW hit in the Red Sea. Although being exceptionally heat resistant, bleaching events in coral reefs in the Red Sea have recently increased in frequency. Genevier et al. (2019) used satellite SST data over the period 1982-2015 to establish Atlas of MHWs hotspots over the Proper Red Sea coral zones (Fig. 6.7). They divided the basin into four main clusters: northern Red Sea (NRS), north central Red Sea (NCRS), south central Red Sea (SCRS) and southern Red Sea (SRS). Between 1985 and 2015 MHWs were not observed across the entire NRS until 1998, but they were common and lasted for a long time afterwards, affecting the entire coastal region. The identified MHW episodes had a high intensity across the region, with temperatures around 1.5 °C above the percentile threshold, and were highest in the northernmost tip, with temperatures around 1.7 °C. The longest MHW period was 12 days. Only two MHWs were detected in the NCRS until 1998, both in small, remote locations. After that time, the cluster shows the most systematic pattern of MHWs, with events that tend to extend across the entire region and last for more than 20 days. MHWs characterize the

SCRS in summer periods and can last for 40 days with temperature increase of 1.2 °C above the normal threshold. The SRS has the slowest increase in summer SST, estimated at 0.14 °C/decade, while having the highest SST in the Red Sea overall. MHWs tend to occur on a regular basis between 13° N and 15° N, although the areas affected remain confined, and their spatial extent, as well as the number of MHW days, appears to reduce over time. Over the period 1982-2015, the most prolonged MHW lasted for 31 days with a temperature increase of about 1.9 °C off Eritrea. The authors linked the MHW incidents to the coral bleaching events recorded in the Red Sea from 1985 to 2015, and they produced a map (Fig. 6.8) to depict both the bleaching location and the SST exceedance over the thresholds during MHW events.

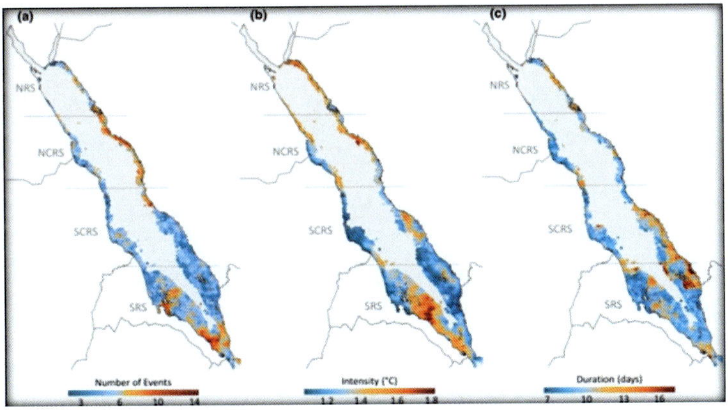

Figure 6.7. Atlas of MHWs over the Red Sea from 1982 to 2015
(a) No. of detected events, (b) increased SST (°C) above thresholds and
(c) MHW period (days) (Source: Genevier et al., 2019)

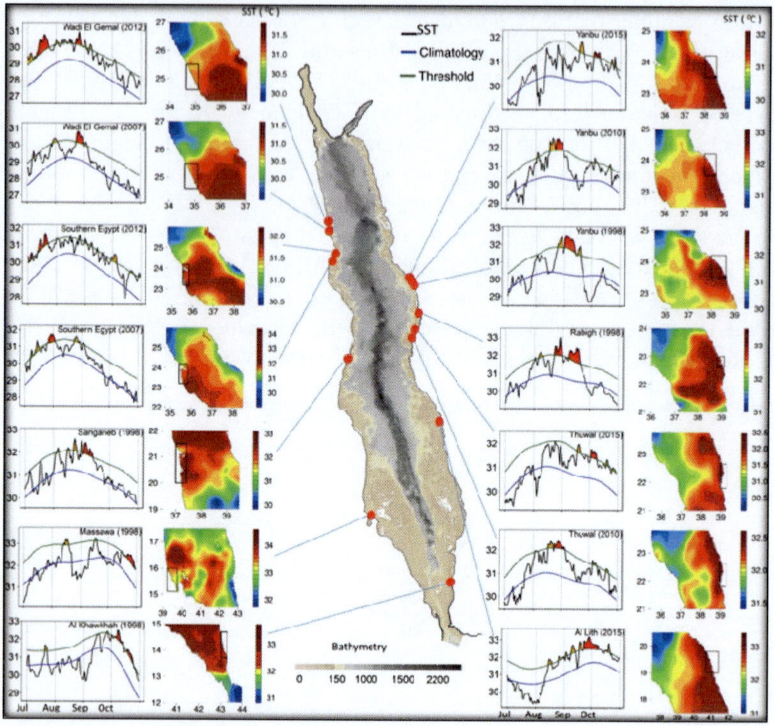

Figure 6.8. Coral regions affected by MHWs in the Red Sea Basin over different periods from 1985 to 2015 (Source: Genevier et al., 2019)

To the best of the author's knowledge, it is necessary to make further research on the event frequency of occurrence, its causes and its destructive impact on the ecosystems of the Red Sea. Coral reefs are not the only ecosystem that can be affected by such extreme event in this important basin. Mangrove swamps, megafauna such as dungeons and sharks, sea birds (local and migrating), seagrasses and fish species are all vital ecosystems tend to be affected. Therefore, the future research on this extreme air-sea interaction event along the Red Sea basin should pay more attention to examine the anthropogenic elements that contribute to MHW (tourism and industry along the Red Sea Coast), and to the meteorological and atmospheric indices impacting the Basin. Management and mitigation plans to conserve natural habitats of the Red Sea should consider the findings of such research; to maintain the safety of the Basin to the most possible extent.

CHAPTER SEVEN

CONCLUSIONS AND RECOMMENDATIONS

To conclude, air-sea interaction processes are natural processes resulting from the interaction between the surface layer of the oceans (~150 m) and the overlying lower layer of the atmosphere (Troposphere). These processes drive and maintain life on Earth. These processes are observed in many dynamical processes, e.g. the global water, heat, and salt balances, the wind system which drives the surface currents around the globe and initiate the surface wave actions, the evaporation and precipitation processes and others. Air-sea interaction processes can be observed on global (wind system, heat balance), regional (monsoons, ENSO) and local (Sea/land breeze, surface waves, surges) scales.

The assessment of the air-sea interaction processes depends mainly on bulk aerodynamics formulas and equations, which comprise the different parameters affecting each process from both sea and atmosphere. With the advances in the technology of measurements, special instruments have been innovated to record essential parameters affecting air-sea interaction processes, e.g. meteorological conditions (wind regime, atmospheric pressure, air temperature, etc.), the different radiation fluxes (insolation, albedo, back radiation, etc.), the natural phenomena (lighting, storms, etc.). Integrated marine buoys and air-sea interaction profilers by which meteorological and oceanic parameters can be measured and recorded have now been used to assess the air sea interaction processes. Beside these fixed-sensor tools, moving tools, like gliders are now value-added equipment in this concern. Also, the application of numerical models has now facilitated the simulation of different air-sea interaction processes and the drawing of the future projections based on many factors such as geography, spatiotemporal scales, emission scenarios, etc. These numerical models can be used on all global, regional and local scale grids.

Due to the anthropogenic activities, which causes excess emission of GHGs to the atmosphere, these natural air-sea interaction processes have shifted to experience abnormal occurrences, which is known as the 'extreme events'. So, an extreme event can refer to the occurrence of a value of a natural phenomenon above or below a threshold value near the upper or lower ends of the range of its observed values. These extremes are presented in so many forms observed all over the globe, e.g. the excess increase in the ocean SST, the heavy precipitation and droughts patterns, the higher rates of evaporation affecting the heat fluxes at the air-sea boundary, the rogue waves, the metotsunamis, the frequent occurrence of marine heatwaves and so many others. Doubtless, changes and shifting in air-sea interaction processes have severe impacts on human and animal health, food security, agricultural productivity, and therefore influence the socioeconomic sector to a considerable extent.

Projections of future occurrence and ocean-atmosphere coupled model assure higher rates of occurrence of these extreme events and that people all over the world will continue to experience their impacts in this and the upcoming century unless serious actions are taken. Therefore, the annual summits of climate change are a must in order to set applied mitigation and adaptation plans and to draw the convenient roadmaps to save our mother Earth.

Recommendations from the present work can be pointed out as follow:

1. The expansion in building global observing and warning systems.
2. The expansion in collaboration plans on regional and global scales aiming at reducing anthropogenic loads on both oceans and atmosphere.
3. Improving energy efficiency and choosing renewable energy over fossil fuels as a source of energy; in order to reduce GHG emissions.
4. Replanting forests and repairing ecosystems that have been harmed; in order to increase the O_2 production and help oceans and atmosphere reducing the amount of absorbed and emitted CO_2.
5. Investigating and creating novel approaches to preventing and managing natural disasters.
6. Creating action plans for climate emergencies.
7. The public awareness is a must, so enhancing people's knowledge and awareness about the impact of such extreme events and how they affect their daily life is highly recommended.
8. In Egypt, enhancement of the monitoring and observing technologies to adequately assess air-sea interaction processes is a must. More interdisciplinary research on the air-sea interaction processes and extreme events is crucial not to be far behind the global level.
9. The exchange of scientific visits through researchers and postgraduate students is an important element in this field.

10. The author calls on local authorities to be more flexible and lenient when it comes to sponsoring research expeditions and obtaining security clearances of needed equipment.

Experiencing extreme air-sea interaction events will last and continue until restrictive rules are set based on a rigid scientific point of view. The drawn plans must be respected and put in force by people from all over the world. The point is that such processes are transcontinental, and cannot be limited to a specified location. However, even with serious actions taken, positive results are not anticipated to be seen in the near future. What has been changed over years and decades will not return to normal overnight, but let us work for the future generations.

In this concern the author believes that the future will seek more collaboration between the countries on both regional and global scales will be set to maintain our planet. Applied technologies to observe and record the various variables involved in the diverse air-sea interaction processes will be more enhanced to have higher accuracies of the recorded data sets. Data exchange between the concerned data centers will be promoted. The use of the coupled numerical models to simulate and anticipate air-sea interaction processes will continue and will be more reliable by enhancing the applied resolutions and parametrization process. The contribution of the satellite imagery will be expanded, especially with those satellites designed and specified for ocean and atmospheric sciences, e.g. Meteosat-8, Meteosat-9, ENVISAT, and others.

If the will exist we can change climate change and resilience back to natural phenomenon associated with the air-sea interaction processes avoiding facing impacts from any extreme event.

REFERENCES

Abbs, D., 1986. Sea-Breeze Interactions along a Concave Coastline in Southern Australia. Month. Weather Review 114, 831–848.

Abbs, D., Physick, W.L., 1992. Sea-breeze oebservation and modelling: a review. Australian Meteorological Magazine 41, 7–19.

Abdel Wahab, M., 2010. Overview on the effect of tourism development on the coastal zone, of Red Sea, Egypt. Egyptian Journal of Aquatic Research 36 (2), 264-277.

Aboobacker, V.M., Shanas, P.R., Alsaafani, M.A., Albarakaati, A.M.A., 2016. Wave energy resource assessment for Red Sea. Reneweable Energy 114, 46-58.

Adloff, F., Somot, S., Sevault, F., Jorda, G., Aznar, R., Deque, M., et al., 2015. Mediterranean Sea response to climate change in an ensemble of twenty first century scenarios. Climate Dynamics 45(9–10), 2775–2802.

Ahrens, C.D., 2018. Essentials of Meteorology: An Invitation to the Atmosphere. 3rd ed. Cengage Learning, Inc, 550pp.

Aiki, H., Takahashi, K., Yamagata, T., 2006. The Red Sea outflow regulated by the Indian monsoon. Continental Shelf Research 26, 1448–1468. https://doi.org/10.1016/j.csr.2006.02.017

Al-Akhaly, I.A., Al-Shwafi, N.A., Al-Safany, M.A., 2020. Study on Hydrographical Properties in the Surface Coastal Waters along the Gulf of Aden and Arabian Sea, Yemen. Advances in Oceanography and Marine Biology 2 (2), 1-8. DOI: 10.33552/AOMB.2020.02.000532.

Al-Barakati, A.M.A., James, A.E., Karakes, G.M., 2002. A three dimensional hydrodyamical model to predict the distribution of temperature, salinity and water circulation of the Red Sea. Journal of King Abdul-Aziz University: Marine Sciences 13, 3-16.

Al-Rousan, S.A., Al-Shloul, R.N., Al-Horani, F.A., Abu-Hilal, A.H., 2007. Heavy metal contents in growth bands of Porites corals: Record of anthropogenic and human developments from the Jordanian Gulf of Aqaba. Marine Pollution Bulletin 54, 1912–1922.

Al-Saafani, M.A., Shenoi, S.S.C., 2007. Water masses in the Gulf of Aden. Journal of Oceanography 63, 1–14. https://doi.org/10.1007/s10872-007-0001-1

Al-Subhi, A., 2012. Estimation of Evaporation Rates in the Southern Red Sea Based on the AVHRR Sea Surface Temperature Data. Journal of King Abdulaziz University-Marine Sciences 23, 77-89. https://doi.org/10.4197/Mar.23-1.6

Al-Subhi, A., Al-Aqsum, M., 2008. Temporal and spatial variations of remotely sensed sea surface temperature in the northern Red Sea. Journal of King Abdul-Aziz University: Marine Sciences 19, 61-74.

Al-Taani, A., Batayneh, A., Mogren, S., Nazzal, Y., Ghrefat, H., Zaman, H., et al., 2013. Groundwater quality of coastal aquifer systems in the eastern coast of the Gulf of Aqaba, Saudi Arabia. Journal of Applied Science and Agriculture 8, 768-778.

Al-Taani, A.A., Rashdan, M., Nazzal, Y., Howari, F., Iqbal, J., Al-Rawabdeh, et al., 2020. Evaluation of the Gulf of Aqaba Coastal Water, Jordan. Water 12, 2125. https://doi.org/10.3390/w12082125

Alawad, K.A., Al-Subhi, A.M., Alsaafani, M.A., Alraddadi, T.M., Ionita, M., Lohmann, G., 2019. Large-Scale Mode Impacts on the Sea Level over the Red Sea and Gulf of Aden. Remote Sensing 11, 2224. https://doi.org/10.3390/rs11192224

Alfartusi, A.J.M., 2019. Estimation of evaporation rates in the north west Arabian gulf based on sea surface temperature and Meteorological data. University of Thi-Qar Journal of Science 7, 37–40. https://doi.org/10.32792/utq/utjsci/vol7/1/7

Alharbi, M.M.E., 2018. Analysis of extreme precipitation events over the eastern Red Sea coast for recent and future climate conditions. Ph.D. Dissertation, University of Birmingham, UK, 211pp.

Alkhalaf, A. K., Basser, H. A., 2013. Diagnostic study of a severe thunderstorm over Jeddah. Atmospheric and Climate Sciences 3, 15.

Anonymus, 2012. Record-breaking La Nina events: An Analysis of the La Nina Life Cycle and the Impacts and Significance of 2010-11 and 2011-12 La Nina Events in Australia. Bureau of Meteorology of Australia, Melbourne, Australia, 24pp.

Arabelos, D.N., Papazachariou, D.Z., Contadakis, M.E., Spatalas, S.D., 2010. A new assimilation tidal model for the Mediterranean Sea. Ocean Science Discussions 7, 1703-1737.

Azorin-Molina, C., Chen, D., 2009. A climatological study of the influence of synoptic-scale flows on sea breeze evolution in the Bay of Alicante (Spain). Theoretical and Applied Climatology 96, 249–260. https://doi.org/10.1007/s00704-008-0028-2

Azorin-Molina, C., Martin-Vide, J., 2007. Methodological approach to the study of the daily persistence of the sea breeze in Alicante (Spain). Atmosfera 20, 57–81.

Azorin-Molina, C., Tijm, S., Chen, D., 2011. Development of selection algorithms and databases for sea breeze studies. Theoretical and Applied Climatology 106, 531–546.

Baldi, M., Amin, D., Al Zayed, I.S., Dalu, G., 2020. Climatology and dynamical evolution of extreme rainfall events in the inai Peninsula—Egypt. Sustainability 12, 6186.

Barale, V., Schiller, C., Villacastin, C., Tacchi, R., 2004. The Adriatic Sea surface temperature historical record from Advanced Very High Resolution Radiometer data (1981–1999). International Journal of Remote Sensing 25, 1363–1370.

Batayneh, A.T., Ghrefat, H., Zumlot, T.T., Elawadi, E., Mogren, S., Zaman, H., et al., 2014. Assessing of metals and metalloids in surface sediments along the Gulf of Aqaba coast, northwestern Saudi Arabia. Journal of Coastal Research 31, 163-176.

Batisha, A.F., 2015. Implementing fuzzy decision making technique in analyzing the Nile Delta resilience to climate change. Alexandria Engineering Journal 54, 1043-1056.

Bechle, A.J., Wu, C.H., Kristovich, D.A.R., Anderson, E.J., Schwab, D.J., Rabinovich, A.B., 2016. Meteotsunamis in the Laurentian Great Lakes. Scientific Reports 6, 37832.

Belkin, I.M, 2009. Rapid warming of large marine ecosystems. Progress in Oceanography 81, 207-213.

Benestad, R., Nychka, D., Mearns, L., 2012. Spatially and temporally consistent prediction of heavy precipitation from mean values. Nature Climate Change 2, 544-547. DOI:10.1038/NCLIMATE1497.

Bengtsson, L., 2010. The global atmospheric water cycle. Environmenatl Research Letters 5, 025202.
https://doi.org/10.1088/1748-9326/5/2/025202

Bernie, D.J., Guilyardi, E., Madec, G., Slingo, J.M., Woolnough, S.J., 2007. Impact of resolving the diurnal cycle in an ocean–atmosphere GCM. Part 1: a diurnally forced OGCM. Climate Dynamics 29, 575–590. https://doi.org/10.1007/s00382-007-0249-6

Bertin, X., Prouteau, E., Letetrel, C., 2013. A significant increase in wave height in the North Atlantic Ocean over the 20th century. Global Planetary Changes 106, 77-83.

Bianchi, C.N., Morri, C., 2000. RMESUaLTrSine biodiversity of the Mediterranean Sea: Situation, problems and prospects for future research. Marine Pollution Bulletin 40, 367-376.

Bigot, S., Planchon, O., 2003. Identification and characterization of sea breeze days in northern France using singular value decomposition. International Journal of Climatology 23, 1397–1405.

Bjerknes, J., 1969. Atmospheric teleconnections from the Equatorial Pacific. Monthly Weather Review 97 (3), 163–172.

Black, E., Blackburn, M., Harrison, G., Hoskins, B., Methven, J., 2004. Factors contributing to the summer 2003 European heatwave. Weather 59, 217-223.

Bourassa, M., Gille, S., Jackson, D., Roberts, J.B., Wick, G., 2010. Ocean Winds and Turbulent Air-Sea Fluxes Inferred From Remote Sensing. Oceanography 23, 36–51.

Breilh, J-F, Bertin, X., Chaumillon, E., Giloy, N., Sauzeau, T., 2014. How frequent is storm-induced flooding in the central part of the Bay of Biscay? Global Planet Change 122, 161-175.

Buckman, L., Verlaan, N., Weerts, A., 2015. Global storm surge forecasting and information system. Geophysical Research Abstracts 17, EGU2015-8323

Butterfield, S., Musial, W., Jonkman, J., Sclavounos, P., 2005. Engineering challenges for floating offshore wind turbines. In: Proceedings Copenhagen Offshore Wind Conference, pp. 377–382.

Carnero, J.A.A., Raya, J.P.B., De la Morena, B.A., 2005. Analysis of sea-land breeze around the City of Huelva (Spain). In: Proceedings 1st International Meeting on Applied Physics (APHYS-2003), pp. 161–165.

Cavaleri, L., Barbariol, F., Bastianini, M., Benetazzo, A., Bertotti, L., Pomaro, A., 2021. An exceptionally high wave at the CNR-ISMAR oceanographic tower in the Northern Adriatic Sea. Scientific Data , 8, 1–7.

Cavaleri, L., Bertotti, L., Torrisi, L., Bitner-Gregersen, E., Serio, M., Onorato, M., 2012. Rogue waves in crossing seas: The Louis Majesty accident. Journal of Geophysical Research 117, C00J10.

Cavicchia, L., von Storch, H., Gualdi, S., 2014. Mediterranean tropical-like cyclones in present and future climate. Journal of Climate 27, 7493-7501.

Chadenas, C., Creach, A., Mercier, D., 2014. The impact of storm Xynthia in 2010 on coastal flood prevention policy in France. Journal of Coastal Conservation 18, 529–538. DOI: 10.1007/s11852-013-0299-3

Chaidez, V., Dreano, D., Agusti, S., Duarte, C.M., Hoteit, I., 2017. Decadal trends in Red Sea maximum surface temperature. Scientific Reports 7 (1), 1–8.

Charnock, H., 1951. Note on eddy diffusion in the atmosphere between one and two kilometres. Quarterly Journal of the Royal Meteorological Society 77, 654–658.

Chen, H.-C., Tseng, Y.-H., Hu, Z.-Z., Ding, R., 2020. Enhancing the ENSO predictability beyond the Spring Barrier. Scientific Reports 10, 984.

Chenbing, Z., Xinpeng, W., Xiyao, L., Suoping, Z., Haitao, W., 2017. A small buoy for flux measurement in air-sea boundary layer. In: Proceedings 13th IEEE International Conference on Electronic Measurement & Instruments (ICEMI), pp. 534–538.

Cheng, L., Trenberth, K.E., Fasullo, J., Boyer, T., Abraham, J., Zhu, J., 2017. Improved estimates of ocean heat content from 1960 to 2015. Science Advances 3, e1601545.

Chowdary, J.S., John, N., Gnanaseelan, C., 2014. Interannual variability of surface air-temperature over India: impact of ENSO and Indian Ocean sea surface temperature. International Journal of Climatology 34, 416–429. https://doi.org/10.1002/joc.3695

Christakos, K., Cheliotis, I., Varlas, G., Steeneveld, G.-J., 2016. Offshore wind energy analysis of cyclone Xaver over North Europe. Energy Procedia 94, 37–44.

Christakos, K., Varlas, G., Reuder, J., Katsafados, P., Papadopoulos, A., 2014. Analysis of a low-level coastal jet off the western coast of Norway. Energy Procedia 53, 162–172.

Churchill, J.H., Bower, A.S., McCorkle, D.C., Abualnaja, Y., 2014. The transport of nutrient-rich Indian Ocean water through the Red Sea and into coastal reef systems. Journal of Marine Research 72, 165–181. https://doi.org/10.1357/002224014814901994

Cid, A., Camus, P., Castanedo, S., Méndez, F. J., Medina, R., 2017. Global reconstructed daily surge levels from the 20th Century reanalysis (1871–2010). Global Planetary Change 148, 9–21.

Cid, A., Menéndez, M., Castanedo, S., Abascal, A.J., Méndez, F.J., Medina, R., 2016. Long-term changes in the frequency, intensity and duration of extreme storm surge events in southern Europe. Climate Dynamics 46 (5), 1502-1516.

Cid, A., Wahl, T., Chambers, D. P., Muis, S., 2018. Storm surge reconstruction and return water level estimation in Southeast Asia for the 20th Century. Journal of Geophysical Research 123, 437–451. DOI: 10.1002/2017JC013143

Coats, S., Smerdon, J.E., Cook, B.I., Seager, R., Cook, E.R., Anchukaitis, K.J., 2016. Internal ocean-atmosphere variability drives megadroughts in Western North America. Geophysical Research Letters 43, 9886-9894.

Collins, M., 2000. Understanding uncertainties in the response of ENSO to greenhouse warming. Geophysical Research Letters 27, 3509–3512. https://doi.org/10.1029/2000GL011747

Constable, A.J., Melbourne-Thomas, J., Corney, S.P., Arrigo, K.R., Barbraud, C., Barnes, D.K.A., et al., 2014. Climate change and Southern Ocean ecosystems I: how changes in physical habitats directly affect marine biota. Global Change Biology 20 (10), 3004–3025.

Conte, D., Lionello, P., 2014. Storm surge distribution along the Mediterranean coast: Characteristics and evolution. Procedia-Social and Behavioral Sciences 120, 110-115.

Cools, M., Moons, E., Wets, G., 2010. Assessing the impact of weather on traffic intensity. Weather, Climate, and Society 2 (1), 60–68. https://doi.org/10.1175/2009WCAS1014.1

Cornish, M., Ives, E., 2009. Reeds Maritime Meteorology. 3rd Edition Adlard Coles Nautical, 266pp.

Cronin, M.F., Gentemann, C.L., Edson, J., Ueki, I., Bourassa, M., Brown, S., et al., 2019. Air-sea fluxes with a focus on heat and momentum. Frontiers in Marine Science 6, 430.

Csanady, G.T., 2004. Air-Sea Interaction: Laws and Mechanisms. Cambridge University Press, UK, 239pp.

Curriero, F.C., Patz, J.A., Rose, J.B., Lele, S., 2001. The association between extreme precipitation and waterborne disease outbreaks in the United States, 1948–1994. American Journal of Public Health 91 (8), 1194–1199. https://doi.org/10.2105/AJPH.91.8.1194

Dabbous, A.S., 2020. Study of sea-wave, near-shore current and their impacts on the Red Sea coast of Egypt. Ph.D. Dissertation, Alexandria University, Egypt, 208pp.

Dangendorf, S., Arns, A., Pinto, J. G., Ludwig, P., Jensen, J., 2016. The exceptional influence of storm 'Xaver' on design water levels in the German bight. Environmental Research Letters 11 (5), 054001.

Darmaraki, S., Somot, S., Sevault, F., Nabat, P., Narvaez, W.D.C., Cavicchia, L., et al., 2019. Future evolution of Marine Heatwaves in the Mediterranean Sea. Climate Dynamics 53, 1371-1392.

Dasari, H.P., Langodan, S., Viswanadhapalli, Y., Vadlamudi, B.R., Papadopoulos, V.P., Hoteit, I., 2017. ENSO influence on the interannual variability of the Red Sea convergence zone and associated rainfall. International Journal of Climatology 38, 761-775.

Davis, S.R., Farrar, J.T., Weller, R.A., Jiang, H., Pratt, L.J., 2019. The land-sea breeze of the Red Sea: Observations, simulations, and relationships to regional moisture transport. Journal of Geophysical Research: Atmospheres 124, 13803–13825. https://doi.org/10.1029/2019JD031007

Denamiel, C., Sepic, J., Ivankovic, D., Vilibic, I., 2019a. The Adriatic Sea and coast modelling suite: Evaluation of the meteotsunami forecast component. Ocean Modeling 135, 71–93.

Denamiel, C., Sepic, J., Huan, X., Bolzer, C., Vilibic, I., 2019b. Stochastic surrogate model for meteotsunami early warning system in the eastern Adriatic Sea. Journal of Geophysical Research 124, 8485–8499. https ://doi.org/10.1029/2019J C0155 74

Deser, C., Phillips, A.S., Alexander, M.A., 2010. Twentieth century tropical sea surface temperature trends revisited. Geophysical Research Letters 37, L10701.

De Vries, A.J., Tyrlis, E., Edry, D., Krichak, S.O., Steil, B., Lelieveld, J., 2013. Extreme precipitation events in the Middle East: Dynamics of the Active Red Sea Trough. Journal of Geophysical Research 118, 7087-7108.

De Vries, A.J., Feldstein, S.B., Tyrlis, E., Riemer, M., Baumgart, M., Fnais, M., et al., 2016. Dynamics of tropical-extratropical interactions and extreme precipitation events in Saudi Arabia in autumn, winter and spring. Quarterly Journal of the Royal Meteorological Society 142, 1862-1880.

Diffenbaugh, N.S., Pal, J.S., Giorgi, F., Gao, X., 2007. Heat stress intensification in the Mediterranean climate change hotspot. Geophysical Research Letters 34 (11), L11706.

Dong, B., Sutton, R.T., Shaffrey, L., Wilcox, L.J., 2020. Attribution of 2012 extreme climate events: does air-sea interaction matter? Climate Dynamics 55, 1225–1245.

Drakopoulos, P.G., Lascaratos, A., 1999. Modelling the Mediterranean Sea: Climatological forcing. Journal of Marine Systems 20, 157–173.

Dugdale, S.J., Hannah, D.M., Malcolm, I.A., 2017. River temperature modelling: A review of process-based approaches and future directions. Earth-Science Reviews 175, 97–113.

Dunsmuir, W.T.M., Spark, E., Kim, S.-K., Shen, S., 2003. Statistical prediction of sea breezes in Sydney Harbour. Australian Meteorology Magazine 52, 117–126.

Dysthe, K., Krogstad, H.E., Müller, P., 2008. Oceanic rogue waves. Annual Review of Fluid Mechanics 40, 287–310. https://doi.org/10.1146/annurev.fluid.40.111406.102203

Eager, R.E., Raman, S., Wootten, A., Westphal, D.L., Reid, J.S., Al Mandoos, A., 2008. A climatological study of the sea and land breezes in the Arabian Gulf region. Journal of Geophysical Research 113, D15106. https://doi.org/10.1029/2007JD009710

Eakin, C.M., Morgan, J.A., Heron, S.F., Smith, T.B., Liu, G., Alvarez-Filip, L., et al., 2010. Caribbean corals in crisis: record thermal stress, bleaching, and mortality in 2005. PLoS One 5, e13969.

Eid, F.M., 1998. Variation of surge heights at Alexandria Port (Egypt). Journal of King Abdul-Aziz University: Marine Sciences 9, 3–18.

El-Isa, Z.H., 2015. Seismicity and seismotectonics of the Red Sea region. Arabian Journal of Geosciences 8, 8505-8525.

El-Geziry, T.M., 2021. Records of surface meteorological elements and their applications in Alexandria Eastern Harbour, Egypt. Athens Journal of Mediterranean Studies 7, 155–170.

El-Geziry, T.M., 2021. Long-term changes in sea surface temperature (SST) within the southern Levantine Basin. Acta Oceanologica Sinica 40, 1–7.

El-Geziry, T.M., Dabbous, A.S., 2021. Behavior of surges in Alexandria Eastern Harbor, Egypt. Egyptian Journal of Aquatic Research 47, 171-178. https://doi.org/10.1016/j.ejar.2021.02.001

El-Geziry, T.M., Dabbous, A.S., Abdallah, A.M., Eid, F.M., 2021. Temporal variability of sea-level in Safaga Harbour, Egypt in relation with meteorological conditions and tidal characteristics. Arabian Journal of Geosciences 14, 1206.

El-Geziry, T.M., Elbessa, M., Tonbol, K.M., 2021. Climatology of sea–land breezes along the southern coast of the Levantine Basin. Pure and Applied Geophysics 178, 1927-1941.

El-Gindy, A.A., Sharf el-Din, S.H. and Moursy, Z.A., 1992. Forecasting of the storm surge in winter at Alexandria (Egypt). Qatar University Science Journal 12, 239-244.

Faizal, M., Ahmed, M.R., 2011. On the ocean heat budget and ocean thermal energy conversion: Ocean heat budget and OTEC. International Journal of Energy Research. 35, 1119–1144.

Field, C.B., Barros, V., Stocker, T.F., Dahe, Q., 2012. Managing the Risks of Extreme Events and Disasters to Advance Climate Change Adaptation: Special Report of the Intergovernmental Panel on Climate Change. Cambridge University Press, Cambridge, UK, 582pp.

Fordyce, A.J., Ainsworth, T.D., Heron, S.F., Leggat, W., 2019. Marine heatwave hotspots in coral reef environments: Physical drivers, ecophysiological outcomes, and impact upon structural complexity. Frontiers in Marine Science 6, 498. doi: 10.3389/fmars.2019.00498

Fortunato, A.B., Li, K., Bertin, X., Rodrigues, M., Miguez, B.M., 2016. Determination of extreme sea levels along the Iberian Atlantic coast. Ocean Engineering 111, 471-482.

Froidurot, S., Diedhiou, A., 2017. Characteristics of wet and dry spells in the West African monsoon system: Characteristics of wet and dry spells in West Africa. Atmospheric Science Letters 18, 125–131. https://doi.org/10.1002/asl.734

Frölicher, T.L., Laufkötter, C., 2018. Emerging risks from marine heat waves. Nature Communications 9, 650.

Frölicher, T.L., Fischer, E.M., Gruber, N., 2018. Marine heatwaves under global warming. Nature 560, 360–364.

Furberg, M., Steyn, D.G., Baldi, M., 2002. The climatology of sea breezes on Sardinia. International Journal of Climatology 22, 917–932. https://doi.org/10.1002/joc.780

Gad, M., Eid, F., Sharaf El-Din, S., Radwan, A., Soliman, G., 2019. Estimation of the salt storage and the salt content in the Gulf of Suez. Journal of King Abdul-Aziz University: Marine sciences 29 (1), 37-51.

Galli, G., Solidoro, C., Lovato, T., 2017. Marine heat waves hazard 3D maps, and the risk for low motility organisms in a warming Mediterranean Sea. Frontiers in Marine Science 4, 136.

Garrabou, J., Coma, R., Bensoussan, N., Bally, M., Chevaldonné, P., Cigliano, M., et al. 2009. Mass mortality in Northwestern Mediterranean rocky benthic communities: Effects of the 2003 heat wave. Global Change Biology 15, 1090–10103.

Geen, R., Bordoni, S., Battisti, D.S., Hui, K., 2020. Monsoons, ITCZs, and the concept of the global monsoon. Reviews of Geophysics 58, 1-45. https://doi.org/10.1029/2020RG000700

Genevier, L.G., Jamil, T., Raitsos, D.E., Krokos, G., Hoteit, I., 2019. Marine heatwaves reveal coral reef zones susceptible to bleaching in the Red Sea. Global Change Biology 25, 2338–2351.

Glozman, M., Sladkevich, M., 2019. Screening of severe storms for possible rogue waves. Computational Technologies 24, 7-15.

Graber, H.C., Terray, E.A., Donelan, M.A., Drennan, W.M., Peters, D.B., 2000. ASIS - A new Air–Sea Interaction Spar buoy: Design and performance at sea. Journal of Atmospheric and Oceanic Technology 17, 708-720.

Granados, R., Soria, J., Cortina, M., 2017. Rainfall variability, rainfed agriculture and degree of human marginality in North Guanajuato, Mexico: Rainfall variability in North Guanajuato. Singapore Journal of Tropical Geography 38, 153–166. https://doi.org/10.1111/sjtg.12191

Grodsky, S.A., Reul, N., Bentamy, A., Vandemark, D., Guimbard, S., 2019. Eastern Mediterranean salinification observed in satellite salinity from SMAP mission. Journal of Marine Systems 198, 103190. https://doi.org/10.1016/j.jmarsys.2019.103190

Groisman, P.Y., Knight, R.W., Easterling, D.R., Karl, T.R., Hegerl, G.C., Razuvaev, V.N., 2005. Trends in intense precipitation in the climate record. Journal of Climate 18, 1326-1350.

Guha-Sapir, D., Hoyois, P., Wallemacq, P., Below, R., 2016. Annual Disaster Statistical Review 2016: The Numbers and Trends. Collaborating Centre for Research on the Epidemiology of Disasters, Université catholique de Louvain – Brussels, Belgium, 80pp.

Gupta, A.S., Thomsen, M., Benthuysen, J.A., Hobday, A.J., Oliver, E., Alexander, L.V., et al., 2020. Drivers and impacts of the most extreme marine heatwave events. Scientific Reports 10, 19359.

Gusiakov, V.K., Dunbar, P.K., Arcos, N., 2019. Twenty-five years (1992–2016) of global tsunamis: Statistical and analytical overview. Natural Hazards 176, 2795–2807.

Haggag, M., El-Badry, H., 2013. Mesoscale numerical study of quasi-stationary convective system over Jeddah in November 2009. Atmospheric and Climate Sciences 3, 73-86.

Haigh, I.D., Macpherson, L.R., Mason, M.S., Crompton, R.P., George, S., 2014. Estimating present day extreme water level exceedance probabilities around the coastline of Australia: Tropical cyclone induced storm surges. Climate Dynamics 42, 139–157.

Halo, I., 2016. The oceans' role in the hydrological cycle, In: Regional State of the Coast Report: Western Indian Ocean. United Nations Environment Programme (UNEP), pp. 181–191.

Hamed, A.A., 1983. Atmospheric circulation over the south eastern part of the Mediterranean. Ph.D. Dissertation, Alexandria University, Egypt, 280pp.

Hamed, A.A., El-Gindy, A.A.,1988. Strom surge generation by winter cyclones at Alexandria, Egypt. International Hydrographic Review LXV, 129-139.

Harper, B., Hardy, T., Mason, L., Bode, L., Young, I., Nielsen, P., 2001. Queensland Climate Change and Community Vulnerability to Tropical Cyclones, Ocean Hazards Assessment. Department of Natural Resources and Mines, Queensland, Brisbane, Australia, 383pp.

Hazra, A., Huser, R., 2021. Estimating high-resolution Red Sea surface temperature hotspots, using a low-rank semiparametric spatial model. The Annals of Applied Statistics 15 (2), 572-596. DOI:10.1214/20-aoas1418

He, H., Xu, Y., 2016. Wind-wave hindcast in the Yellow Sea and the Bohai Sea from the year 1988 to 2002. Acta Oceanologica Sinica 35, 46–53. https://doi.org/10.1007/s13131-015-0786-5

He, H., Wu, Q., Chen, D., Sun, J., Liang, C., Jin, W., Xu, Y., 2018. Effects of surface waves and sea spray on air–sea fluxes during the passage of Typhoon Hagupit. Acta Oceanologica Sinica 37, 1–7. https://doi.org/10.1007/s13131-018-1208-2

He, Z., Wu, R., Wang, W., Wen, Z., Wang, D., 2017. Contributions of surface heat fluxes and oceanic processes to tropical SST changes: Seasonal and regional dependence. Journal of Climate 30, 4185–4205. https://doi.org/10.1175/JCLI-D-16-0500.1

Head, S., EdwardsA., 1987. Red Sea Key Environments. Pergamon Press, 441pp.

Hendy, D.M., El-Geziry, T.M., El Raey, M., Nasr, S.M., 2021. Sea level characteristics and extremes along Alexandria coastal zone. Arabian Journal of Geosciences 14 (13), 1273.

Hicks, S.D., Sillcox, R.L., Nichols, C.R., Via, B., McCray, E.C., 2000. Tide and Current Glossary. National Ocean Service, USA, 29pp.

Hobday, A.J., Alexander, L.V., Perkins, S.E., Smale, D.A., Straub, S.C., Oliver, E.C., et al., 2016. A hierarchical approach to defining marine heatwaves. Progress in Oceanography 141, 227–238. DOI: 10.1016/j.pocean.2015.12.014

Hobday, A.J., Oliver, E.C., Gupta, A., Benthuysen, J.A., Burrows, M.T., Donat, M.G., et al., 2018. Categorizing and naming marine heatwaves. Oceanography 31, 162–173.

Hoegh-Guldberg, O., Cai, R., Poloczanska, E.S., Brewer, P.G., Sundby, S., Hilmi, K., et al., 2014. The ocean - supplementary material. In: Barros, V.R., Field, C.B., Dokken, D.J., Mastrandrea, M.D., Mach, K.J., Bilir, T.E., et al. (Eds.), Climate Change 2014: Impacts, Adaptation, and Vulnerability. Part B: Regional Aspects. Contribution of Working Group II to the Fifth Assessment Report of the Intergovernmental Panel on Climate Change. Cambridge University Press, Cambridge, pp.1655-1731.

Holbrook, N.J., Scannell, H.A., Gupta, A.S., Benthuysen, J.A., Feng, M., Oliver, E.C., et al. 2019. A global assessment of marine heatwaves and their drivers. Nature Communications 10, 2624

Holding, T., Ashton, I.G., Shutler, J.D., Land, P.E., Nightingale, P.D., Rees, A.P., et al., 2019. The flux engine air–sea gas flux toolbox: Simplified interface and extensions for in situ analyses and multiple sparingly soluble gases. Ocean Sciences 15, 1707–1728.

Holthuijsen, L.H., Powell, M.D., Pietrzak, J.D., 2012. Wind and waves in extreme hurricanes. Journal of Geophysical Research 117, C09003. https://doi.org/10.1029/2012JC007983

Horvat, M., Kotnik, J., Logar, M., Fajon, V., Zvonaric, T., Pirrone, N., 2003. Speciation of mercury in surface and deep-sea waters in the Mediterranean Sea. Atmospheric Environment 37, 93–108. https://doi.org/10.1016/S1352-2310(03)00249-8

Houghton, J.T., Ding, Y., Griggs, D.J., Monguer, M., van der Linden, P.J., Dai, X., et al., 2001. Climate Change 2001: The Scientific Basis. Contribution of the WGI to the 3rd Assessment Report of the Intergovernmental Panel on Climate Change (IPCC). Cambridge University Press, UK, 881pp.

Hussein, M.A., El Geziry, T., 2014. Diurnal variability of heat fluxes and Bowen ratio over Alexandria Eastern Harbor, Egypt. Journal of King Abdulaziz University: Marine Sciences 25, 57-85. https://doi.org/10.4197/Mar.25-2.4

Ibrahim, H.D., Xue, P., Eltahir, E.A.B., 2020. Multiple salinity equilibria and resilience of Persian/Arabian Gulf basin salinity to brine discharge. Frontiers in Marine Science 7, 573. https://doi.org/10.3389/fmars.2020.00573

Ibrahim, O., 2021. Study of the sea level variability and storm surges at Mersa Matruh, Egypt. Arabian Journal of Geosciences 14, 1678.

Ibrahim, O., Mohamed, B., Nagy, H., 2021. Spatial variability and trends of marine heatwaves in the Eastern Mediterranean Sea over 39 Years. Journal of Marine Science and Engineering 9, 643.

Idier, D., Dumas, F., Muller, H., 2012. Tide-surge interaction in the English Channel. Natural Hazard Earth System 12, 3709-3718.

Idier, D., Bertin, X., Thompson, P., Pickering, M.D., 2020. Interactions between mean sea level, tide, surge, waves and flooding: Mechanisms and contributions to sea level variations at the coast. Surveys in Geophysics 40, 1603–1630. https://doi.org/10.1007/s10712-019-09549-5

Innocenti, A., Onorato, M., Brandini, C., 2021. Analysis of dangerous sea states in the Northwestern Mediterranean area. Journal of Marine Science and Engineering 9, 422.

Ionita, M., Felis, T., Lohmann, G., Rimbu, N., Pätzold, J., 2014. Distinct modes of East Asian winter monsoon documented by a southern Red Sea coral record. Journal of Geophysical Research 119, 1517–1533. https://doi.org/10.1002/2013JC009203

Intergovernmental Panel on Climate Change (IPCC) 2007. Climate Change: The Physical Science Basis. Solomon, S., Dahe, Q., Manning,

M., Chen, Z., Marquis, M., Averyt, K. B., Tignor, M., and Miller, H. L. (Eds.) Cambridge University Press, 996p.

Janssen, P., Alpers, W., 2006. Why SAR wave mode data of ERS and ENVISAT are inadequate for giving the probability of occurrence of freak waves. In: Proceedings of SEASAR Conference, pp. 1-11.

Janssen, P., Breivik, O., Mogensen, K., Vitart, F., Balmaseda, M., Bidlot, J.R., et al., 2013. Air-Sea Interaction and Surface Waves (Report No. 712), ECMWF Technical Memoranda, 34pp.

Jiang, H., Farrar, J.T., Beardsley, R.C., Chen, R., Chen, C., 2009. Zonal surface wind jets across the Red Sea due to mountain gap forcing along both sides of the Red Sea. Geophysical Research Letters 36, L19605.

Kataoka, T., Tozuka, T., Yamagata, T. 2017. Generation and decay mechanisms of Ningaloo Nino/Nina. Journal of Geophysical Research 122, 8913–32.

Katzenberger, A., Schewe, J., Pongratz, J., Levermann, A., 2020. Robust increase of Indian monsoon rainfall and its variability under future warming in CMIP-6 models. Earth System Dynamics 12, 367-386. https://doi.org/10.5194/esd-2020-80

Kawamura, H., Qin, H., Ando, K., 2008. *In-situ* diurnal sea surface temperature variations and near-surface thermal structure in the tropical hot event of the Indo-Pacific warm pool. Journal of Oceanography 64, 847–857. https://doi.org/10.1007/s10872-008-0070-9

Kelley, C.P., Mohtadi, S., Cane, M.A., Seager, R., Kushnir, Y., 2015. Climate change in the Fertile Crescent and implications of the recent Syrian drought. Proceedings of National Academy of Sciences 112, 3241-3246.

Kettle, H., Merchant, C.J., Jeffery, C.D., Filipiak, M.J., Gentemann, C.L., 2009. The impact of diurnal variability in sea surface temperature on the central Atlantic air-sea CO_2 flux. Atmospheric Chemistry and Physics 9, 529–541. https://doi.org/10.5194/acp-9-529-2009

Kharif, C., Pelinovskii, E.N., Slunyaev, A., 2009. Rogue waves in the ocean, In: Advances in Geophysical and Environmental Mechanics and Mathematics, 1st Ed. Springer, Berlin, 216pp.

Kim, S., 2019. Storm surges, In: Encyclopedia of Ocean Sciences. Elsevier, pp. 663–671. https://doi.org/10.1016/B978-0-12-409548-9.10849-8

Knapp, A.K., Beier, C., Briske, D.D., Classen, A.T., Luo, Y., Reichstein, M., et al., 2008. Consequences of more extreme precipitation regimes for terrestrial ecosystems. Bioscience 58 (9), 811–821. https://doi.org/10.1641/B580908

Knobler, S., Bar, D., Cohen, R., Liberzon, D., 2021. Wave height distributions and rogue waves in the Eastern Mediterranean. Journal of Marine Science and Engineering 9, 660.

Kontogianni, A.V., Bartzokas, A., 2005. A study on evaporation in Ioannina, NW Greece. In: Proceedings 9th International Conference on Environmental Science and Technology, pp. B443-B448.

Krestinitis, Y.N., Androulidakis, I., Kontos, Y., Georgakopoulos, G., 2011. Coastal inundation in the north-eastern Mediterranean coastal zone due to storm surge events. Journal of Coastal Conservation 15, 353-368.

Krichak, S.O., Breitgand, J.S., Feldstein, S.B., 2012. A conceptual model for the identification of the Active Red Sea Trough Synoptic events over the southeastern Mediterranean. Journal of Applied Meteorology and Climatology 51 (5), 962-971.

Landerer, F.W., Jungclaus, J.H., Marotzke, J., 2008. El Nino–Southern Oscillation signals in sea level, surface mass redistribution, and degree-two geoid coefficients. Journal of Geophysical Research 113, C08014. https://doi.org/10.1029/2008JC004767

Langodan, S., Cavaleri, L., Viswanadhapalli, Y., Hoteit, I., 2014. The Red Sea: a natural laboratory for wind and wave modeling. Journal of Physical Oceanography 44, 3139-3159.

Langodan, S., Cavaleri, L., Pomaro, A., Viswanadhapalli, Y., Bertotti, L., Hoteit, I., 2017. The climatology of the Red Sea - part 2: the waves. International Journal of Climatology 37, 4518-1528. http://dx.doi.org/10.1002/joc.5101.

Lau, K., 1985. Elements of a stochastic-dynamical theory of the long-term variability of the El Nino Southern Oscillation. Journal of the Atmospheric Sciences 42, 1552–1558.

Lau, W.K.M., Waliser, D.E., 2012. Intraseasonal Variability in the Atmosphere-Ocean Climate System. Springer Berlin Heidelberg, Berlin, 613pp. https://doi.org/10.1007/978-3-642-13914-7

Lau, W.K.M., Kim, K.M., Lee, M.I., 2007. Characteristics of diurnal and seasonal cycles in global monsoon systems. Journal of the Meteorological Society of Japan 85A, 403–416.

Lawal, K.A., 2015. Understanding the variability and predictability of seasonal climates over West and Southern Africa using climate models. Ph.D. Dissertation, University of Cape Town, South Africa, 241pp.

Lewandowska, A.M., Boyce, D.G., Hofmann, M., Matthienssen, B., Sommer, U., Worm, B., 2014. Effects of sea surface warming on marine plankton. Ecology Letters 17, 614-623.

Li, X., Hu, Z.-Z., Huang, B., 2020. Subannual to interannual variabilities of SST in the North Atlantic Ocean. Journal of Climate 33, 5547–5564. https://doi.org/10.1175/JCLI-D-19-0556.1

Madah, F., Mayerle, R., Bruss, G., Bento, J., 2015. Characteristics of tides in the Red Sea region, a numerical model study. Open Journal of Marine Science 5, 193-209.

Mahmoud, M.G., Abul-Khir, E., Ebeid, M., Mohamed, L.A., Fahmy, M.A., Shaban, Kh.S., 2020. An assessment on the coastal seawater quality of the Gulf of Suez, Egypt. Journal of Environmental Protection 11, 34-47.

Mahongo, Sh., 2016. Sea/Air Interaction, In: Regional State of the Coast Report: Western Indian Ocean. United Nations Environment Programme (UNEP), pp. 198–210.

Maiyza, I.A., 1988. Evaporation of coastal water in the NW Red Sea. Bulletin of Institute of Oceanography and Fisheries 14 (1), 75-80.

Makris, Ch., Androulidakis, Y; Krestinitis, Y., Kombiadou, K., Baltikas, V., 2015. Numerical modelling of storm surges in the Mediterranean Sea under climate change. In: Proceedings 36th IAHR World Congress, pp. 1-13.

Manasrah, R., Hasanean, H.M., Al-Rousan, S., 2009. Spatial and seasonal variations of sea level in the Red Sea, 1958–2001. Ocean Science Journal 44, 145–159.

Marcos, M., Tsimplis, M.N., Shaw, A.G.P., 2009. Sea level extremes in southern Europe. Journal of Geophysical Research 114, C01007.

Martelloni, G., Segoni, S., Fanti, R., Catani, F., 2012. Rainfall thresholds for the forecasting of landslide occurrence at regional scale. Landslides 9 (4), 485–495.

Masouleh, Z.P., 2015. Identification of Sea Breezes, their Climatic Trends and Causation, with Application to the Adelaide Coast. Ph.D. Dissertation, University of Adelaide, Australia, 200pp.

Masson, V., Zhai, P., Pirani, A., Connors, S.L., Pean, C., Berger, S., et al., 2021. IPCC 2021: Summary for Policymakers, In: Climate Change 2021: The Physical Science Basis. Contribution of Working Group I to the Sixth Assessment Report of the Intergovernmental Panel on Climate Change. Cambridge University Press, 42pp.

Melville, W.K., 1996. The Role of surface-wave breaking in air-sea interaction. Annual Review of Fluid Mechanics 28, 279–321.

Merrill, R.T., Gray, P.I., 1982. A Comparison of Large and Small Scale Tropical Cyclones (Report No. 32). Clorado State University, USA, 75pp.

Miller, S., Keim, B., Talbot, R., Mao, H., 2003. Sea breeze: Structure, forecasting, and impacts. Reviews of Geophysics 41, 1011. https://doi.org/10.1029/2003RG000124

Millot, C., Taupier-Letage, I., 2005. Circulation in the Mediterranean Sea, In: Saliot, A. (Ed.), The Mediterranean Sea, Handbook of Environmental Chemistry. Springer Berlin Heidelberg, Berlin, Heidelberg, pp. 29–66. https://doi.org/10.1007/b107143

Mohamed, B., Nagy, H., Ibrahim, O., 2021. Spatiotemporal variability and trends of marine heatwaves in the Red Sea over 38 Years. Journal of Marine Science and Engineering 9, 842.

Mohamed, B., Abdallah, A.M., Alam El-Din, K., Nagy, H., Shaltout, M., 2019. Inter-annual variability and trends of sea level and sea surface temperature in the Mediterranean Sea over the last 25 Years. Pure and Applied Geophysics 176, 3787–3810.

Moisan, J.R., Niiler, P.P., 1998. The seasonal heat budget of theNorth Pacific: Net heat flux and heat storage rates (1950–1990). Journal of Physical Oceanography 28, 401–21.

Monserrat, S., Vilibic, I., Rabinovich, A.B., 2006. Meteotsunamis: Atmospherically induced destructive ocean waves in the tsunami frequency band. Natural Hazards and Earth System Sciences 6, 1035–1051. https ://doi.org/10.5194/nhess -6-1035-2006

Montblanc, T.F., Vousdoukas, M.I., Breyiannis, G., Feyen, L., Salamon, P., Ciavola, P., et al., 2019. Towards robust pan-European storm surge forecasting. Ocean Modelling 133, 129–144.

Morcos, S.A., 1970. Physical and chemical oceanography of the Red Sea. Oceanography and Marine Biology Annual Review 8, 73-202.

Morcos, S.A., AbdAllah, A.M., 2012. Oceanography of the Gulf of Aden. Egyptian Journal of Aquatic Research 38, 77–91. https://doi.org/10.1016/j.ejar.2012.12.001

Moulin, A., 2015. Air-sea interaction at the synoptic- and the meso-scale. Ph.D. Dissertation, Université Grenoble Alpes, France, 109pp.

Moursy, Z.A., 1976. Storm surges along the Alexandria coasts. M.Sc. Thesis, Alexandria University, 151pp.

Mölders, N., Kramm, G., 2014. Lectures in Meteorology (Springer Atmospheric Sciences). Springer International Publishing, Cham, 591pp. https://doi.org/10.1007/978-3-319-02144-7

Muis, S., Verlaan, M., Winsemius, H.C., Aerts, J.C., Ward, P.J., 2016. A global reanalysis of storm surges and extreme sea levels. Nature Communications 7, 1–11.

Myhre, G., Samset, B.H., Hodnebrog, Ø., Andrews, T., Boucher, O., Faluvegi, G., Fläschner, D., et al., 2018. Sensible heat has significantly

affected the global hydrological cycle over the historical period. Nature Communications 9, 1922. https://doi.org/10.1038/s41467-018-04307-4
Naor, R., Potcher, O., Shafir, H., Alpert, P., 2017. An observational study of the summer Mediterranean sea-breeze-front penetration into the complex topography of the Jordan-Rift-Valley. Theoretical and Applied Climatology 127, 275–284.
Nicholls, R.J., Hanson, S., Herweijer, C., Patmore, N., Hallegatte, S., Corfee-Morlot, J., et al., 2008. Ranking port cities with high exposure and vulnerability to climate extremes-exposure estimates. Environmental Working Paper No.1, Paris Organisation for Economic Co-operation and Development (OECD), 62 pp.
Nidheesh, A.G., Lengaigne, M., Vialard, J., Izumo, T., Unnikrishnan, A.S., Meyssignac, B., et al., 2017. Robustness of observation-based decadal sea level variability in the Indo-Pacific Ocean: Robustness of decadal sea level. Geophysical Research Letters 44, 7391–7400.
Nissen, K.M., Ulbrich, U., 2017. Increasing frequencies and changing characteristics of heavy precipitation events threatening infrastructure in Europe under climate change. Natural Hazards and Earth System Sciences 17 (7), 1177–1190.
https://doi.org/10.5194/nhess-17-1177-2017
Nott, J., Green, C., Townsend, I., Callaghan, J., 2014. The world record storm surge and the most intense Southern Hemisphere tropical cyclone: New evidence and modeling. Bulletin of American Meteorological Society 95, 757-765. https://doi.org/10.1175/BAMSD-12-00233.1.
Obeysekera, J., Park, J., 2013. Scenario-based projections of extreme sea levels. Journal of Coastal Research 29 (1), 1-7.
Oddo, P., Guarnieri, A., 2011. A study of the hydrographic conditions in the Adriatic Sea from numerical modelling and direct observations (2000–2008). Ocean Science 7, 549–567.
Okal, E.A., Visser, J.N.J., de Beer, C.H., 2014. The Dwarskersbos, South Africa local tsunami of August 27, 1969: field survey and simulation as a meteorological event. Natural Hazards 74, 251–268. https ://doi.org/10.1007/s1106 9-014-1205-5
Olita, A., Sorgente, R., Natale, S., Ribotti, A., Bonanno, A., Patti, B., 2007. Effects of the 2003 European heatwave on the Central Mediterranean Sea: Surface fluxes and the dynamical response. Ocean Science 3(2), 273–289.
Oliver, J.E., 2005. Encyclopedia of World Climatology. Springer, Dordrecht, the Netherlands, 854pp.

Oliver, E.C.J., Benthuysen, J.A., Darmaraki, S., Donat, M.G., Hobday, A.J., Holbrook, N.J., et al., 2020. Marine heatwaves. Annual review of Marine Science 13, 20.1-20.30.

Oliver, E.C.J., Burrows, M.T., Donat, M.G., Gupta, A., Alexander, L.V., Perkins-Kirkpatrick, S.E., et al., 2019. Projected marine heatwaves in the 21st century and the potential for ecological impact. Frontiers in Marine Science 6, 734.

Oliver, E.C.J., Donat, M.G., Burrows, M.T., Moore, P.J., Smale, D.A., Alexander, L.V., et al. 2018. Longer and more frequent marine heatwaves over the past century. Nature Communications 9, 1324.

Orejarena-Rondón, A.F., Otero-Díaz, L.J., Restrepo L., J.C., Ramos De la Hoz, I.M., Marriaga-Rocha, L., 2018. Methodology for determining the mean and extreme sea level regimes (astronomical and meteorological tides) considering scarce records in microtidal zones: Colombian Caribbean case. DYNA 85, 274–283. https://doi.org/10.15446/dyna.v85n205.70677

Papalexiou, S.M., Montanari, A., 2019. Global and regional increase of precipitation extremes under global warming. Water Resources Research 55, 4901–4914.

Parker, J.K., McIntyre, D., Noble, R.T., 2010. Characterizing fecal contamination in stormwater runoff in coastal North Carolina, USA. Water Research 44 (14), 4186–4194.

Pastor, F., Valiente, J.A., Khodayar, S., 2020. A warming Mediterranean: 38 years of increasing sea surface temperature. Remote Sensing 12, 2687.

Pastor, F., Valiente, J.A., Palau, J.L., 2018. Sea surface temperature in the Mediterranean: Trends and spatial patterns (1982–2016). Pure and Applied Geophysics 175, 4017–4029.

Pattiaratchi, C., Hetzel, Y., 2020. State and trends of Australia's ocean report: Sea surface temperature variability. Integrated Marine Observing Systems Australia, 1.2.1-1.2.4.

Pattiaratchi, C.B., Wijeratne, E.M.S., 2015. Are meteotsunamis an underrated hazard? Philosophical Transactions of the Royal Society A 373, 20140377.

Payne, R.E., 1972. Albedo of the sea surface. Journal of the Atmospheric Sciences 29, 959–970.

Pedreira, A.L., Querino, C.A.S., Biudes, M.S., Machado, N.G., Santos, L.O.F., Ivo, I.O., 2020. Influence of El Nino and La Nina phenomena on seasonality of the relative frequency of rainfall in southern Amazonas mesoregion. Brazilian Journal of Water Resources 25, e24.

Pelinovsky, E., Slunyaev, A., Lopatoukhin, L., Divinsky, B., Levin, B., 2004. Freak wave event in the Black Sea: Observation and modeling. Doklady Earth Sciences 395, 438–443.

Perini, L., Calabrese, L., Salerno, G., Ciavola, P. and Armaroli, C., 2015. Evaluation of coastal vulnerability to flooding: comparison of two different methodologies adopted by the Emilia-Romagna Region (Italy). Natural Hazards and Earth System Science 3, 4315-4352.

Pinet, P.R., 2009. Invitation to Oceanography. 5th Edition, Jones and Bartlett Publishers, Sudbury, 625pp.

Pirazzoli, P.A., 2000. Surges, atmospheric pressure and wind change and flooding probability on the Atlantic coast of France. Oceanologica Acta 23, 643–661. https://doi.org/10.1016/S0399-1784(00)00122-5

Pokhrel, R., Lee, H., 2011. Estimation of the effective zone of sea/land breeze in a coastal area. Atmospheric Pollution Research 2, 106–115. https://doi.org/10.5094/APR.2011.013

Poulos, S.E., Drakopoulos, P.G., Collins, M.B., 1997. Seasonal variability in sea surface oceanographic conditions in the Aegean Sea (Eastern Mediterranean): An overview. Journal of Marine Systems 13, 225–244. https://doi.org/10.1016/S0924-7963(96)00113-3

Probstand, P., Franchello, G., 2012. Global Storm Surge Forecast and Inundation Modelling. Joint Research Centre Scientific and Technical Reports, Publications Office of the European Union, Italy, 47pp.

Prtenjak, M.T., 2003. Main characteristics of sea/land breezes along the eastern coast of the Northern Adriatic. Geofizika 20, 75–92.

Rabinovich, A.B., Vilibic, I., Tinti, S., 2009. Meteorological tsunamis: Atmospherically induced destructive ocean waves in the tsunami frequency band. Physics and Chemistry of the Earth 34, 891–893. https ://doi.org/10.1016/j.pce.2009.10.006

Radovic, V., Iglesias, I., 2019. Extreme weather events: Definition, classification and guidelines towards vulnerability reduction and adaptation management, In: Climate Action. Encyclopedia of the UN Sustainable Development Goals. Springer Nature, Switzerland, pp. 1–13.

Rafiq, S., Pattiaratchi, C., Janeković, I., 2020. Dynamics of the land-sea breeze system and the surface current response in South-West Australia. Journal of Marine Science and Engineering 8, 931. https://doi.org/10.3390/jmse8110931

Rahul, S., Gnanaseelan, C., 2016. Can large scale surface circulation changes modulate the sea surface warming pattern in the Tropical Indian Ocean? Climate Dynamics 46, 3617–3632.

Raitsos, D.E., Pradhan, Y., Brewin, R.J.W., Stenchikov, G., Hoteit, I., 2013. Remote sensing the phytoplankton seasonal succession of the Red Sea. PLoS ONE 8, e64909.

Raitsos, D.E., Hoteit, I., Prihartato, P.K., Chronis, T., Triantafyllou, G., Abualnaja, Y., 2011. Abrupt warming of the Red Sea. Geophysical Research Letters 38, L14601.

Raitsos, D.E., Pradhan, Y., Brewin, R.J.W., Stenchikov, G., Hoteit, I., 2013. Remote sensing the phytoplankton seasonal succession of the Red Sea. PloS ONE 8, e64909.

Raitsos, D.E., Yi, X., Platt, T., Racault, M., Brewin, R.J.W., Pradhan, Y., et al., 2015. Monsoon oscillations regulate fertility of the Red Sea. Geophysical Research Letters 42, 855–862.

Ralston, D.K., Jiang, H., Farrar, J.T., 2013. Waves in the Red Sea: Response to monsoonal and mountain gap winds. Continental Shelf research 65, 1–13.

Ramu, C.V., Bharathi, G., Sadhuram, Y., 2010. Diurnal variability of heat fluxes over the coastal waters off Visakhapatnam during post-monsoon and winter seasons. Indian Journal of Marine Sciences 39, 128–135.

Rasul, N.M.A., Stweart, I.C.F., 2015. The Red Sea: The Formation, Morphology, Oceanography and Environment of a Young Ocean Basin. Springer, 368pp.

Reed, R.K., 1977. On estimating insolation over the ocean. Journal of Physical Oceanography 7, 482–485.

Ren, H.-L., Zheng, F., Luo, J.-J., Wang, R., Liu, M., Zhang, W., et al., 2020. A review of research on tropical air-sea Interaction, ENSO Dynamics, and ENSO prediction in China. Journal of Meteorological Research 34, 43–62.

Renault, L., Vizoso, G., Jansa, A., Wilkin, J., Tintore, J., 2011. Toward the predictability of meteotsunamis in the Balearic Sea using regional nested atmosphere and ocean models. Geophysical Research Letters 38, L10601. https ://doi.org/10.1029/2011G L0473 61

Rogers, D.P., 1995. Air-sea interaction: Connecting the ocean and atmosphere. Reviews of Geophysics 33, 1377–1383. https://doi.org/10.1029/95RG00255

Rohli, R.V., Li, C., 2021. Meteorology for Coastal Scientists. Springer International Publishing, Cham, 525pp. https://doi.org/10.1007/978-3-030-73093-2

Rohrbeck, C., Simpson, E.S., Towe, R.P., 2021. A spatio-temporal model for Red Sea surface temperature anomalies. Extremes 24, 129–144. https://doi.org/10.1007/s10687-020-00383-2

Rosenzweig, C., Tubiello, F.N., Goldberg, R., Mills, E., Bloomfield, J., 2002. Increased crop damage in the US from excess precipitation under climate change. Global Environmental Change 12 (3), 197–202. https://doi.org/10.1016/S0959-3780(02)00008-0

Roxy, M.K., Gnanaseelan, C., Parekh, A., Chowdary, J.S., Singh, S., Modi, A., et al., 2020. Indian Ocean warming, In: Krishnan, R., Sanjay, J., Gnanaseelan, Mujumdar, M., Kulkarni, A., Chakraborty, S. (Eds.), Assessment of Climate Change over the Indian Region. Springer Singapore, Singapore, pp. 191–206. https://doi.org/10.1007/978-981-15-4327-2_10

Russo, A., Artegiani, A., 1996. Adriatic Sea hydrography. Scientia Marina 60, 33–43.

Rutgersson, A., Pettersson, H., Nilsson, E., Bergström, H., Wallin, M.B., Nilsson, E.D., et al., 2020. Using land-based stations for air–sea interaction studies. Tellus A: Dynamic Meteorology and Oceanography 72, 1–23. https://doi.org/10.1080/16000870.2019.1697601

Saad, N., Moursy, Z., Sharaf El-Din, S.H., 2011. Water heights and weather regimes at Alexandria Harbor. International Journal of Physical Sciences 6, 7035–7043.

Said, M.A., El-Geziry, T.M., Radwan, A.A., 2012. Long-term trends of extreme climate events over Alexandria region, Egypt. In: Proceedings of the Land-Sea interactions in coastal zone (LANDSI) Conference, pp. 286-293.

Salaree, A., Mansouri, R., Okal, E.A., 2018. The intriguing tsunami of 19 March 2017 at Bandar Dayyer, Iran: field survey and simulations. Natural Hazards 90, 1277–1307.

Sauvage, C., Lebeaupin Brossier, C., Bouin, M.-N., Ducrocq, V., 2020. Characterization of the air–sea exchange mechanisms during a Mediterranean heavy precipitation event using realistic sea state modelling. Atmospheric Chemistry and Physics 20, 1675–1699.

Schmidt, F.H., 1947. An elementary theory of the land-sea breeze circulation. Journal of Meteorology 4, 9–20.

Schmidt, H., Czeschel, R., Visbeck, M., 2020. Seasonal variability of the circulation in the Arabian Sea at intermediate depth and its link to the Oxygen Minimum Zone. Ocean Science Discussions 16 (6), 1459-1474. https://doi.org/10.5194/os-2020-9

Schott, F.A., McCreary, J.P., 2001. The monsoon circulation of the Indian Ocean. Progress in Oceanography 51, 1–123. https://doi.org/10.1016/S0079-6611(01)00083-0

Schulz, W.J., 2017. National Hurricane Operations Plan (Report No. FCM-P12-2017). Washington, D.C., USA, 185pp.

Sepic, J., Rabinovich, A.B., Sytov, V.N., 2018. Odessa tsunami of 27 June 2014: Observations and numerical modelling. Pure and Applied Geophysics 175, 1545–1572.

Shaltout, M., 2019. Recent sea surface temperature trends and future scenarios for the Red Sea. Oceanologia 61, 484–504. https://doi.org/10.1016/j.oceano.2019.05.002

Shaltout, M., Omstedt, A., 2014. Recent sea surface temperature trends and future scenarios for the Mediterranean Sea. Oceanologia 56, 411–443. https://doi.org/10.5697/oc.56-3.411

Shanas, P.R., Aboobacker, V.M., Albarakati, A.M., Zubier K.M., 2017. Climate driven variability of wind-waves in the Red Sea. Ocean Modelling 119, 105–117.

Shinoda, T., 2005. Impact of the diurnal cycle of solar radiation on intraseasonal SST variability in the Western Equatorial Pacific. Journal of Climate 18, 2628–2636.

Shinoda, T., Zamudio, L., Guo, Y., Metzger, E.J., Fairall, C.W., 2019. Ocean variability and air-sea fluxes produced by atmospheric rivers. Scientific Reports 9, 2152.

Shultz, J.M., Russell, J., Espinel, Z., 2005. Epidemiology of tropical cyclones: Dynamics of disaster, disease, and development. Epidemiologic Reviews 27, 21–35.

Skliris, N., Sofianos, S., Gkanasos, A., Mantziafou, A., Vervatis, V., Axaopoulos, P., et al., 2012. Decadal scale variability of sea surface temperature in the Mediterranean Sea in relation to atmospheric variability. Ocean Dynamics 62, 13–30. https://doi.org/10.1007/s10236-011-0493-5

Slingo, J., Spencer, H., Koskins, B., Berrisford, P., Black, E., 2005. The meteorology of the Western Indian Ocean and the influence of the East African Highlands. Philosophical Transactions of the Royal Society A: Mathematical, Physical and Engineering Sciences 363, 1826.

Smale, D.A., Wernberg, T., Oliver, E.C.J., Thomsen, M., Harvey, B.P., Straub, S.C., et al., 2019. Marine heatwaves threaten global biodiversity and the provision of ecosystem services. Nature Climate Change 9, 306–312.

Smith, C.B., 2007. Extreme Waves and Ship Design. In: Proceedings 10th International Symposium on Practical Design of Ships and Other Floating Structures, pp. 1-8.

Sobel, A.H., Tippett, M.K., 2018. Extreme events, In: Resilience. Elsevier, pp. 3–12.

Sofianos, S.S., Johns, W.E., 2003. An Oceanic General Circulation Model (OGCM) investigation of the Red Sea circulation: 2. Three-dimensional circulation in the Red Sea. Journal of Geophysical Research 108, 3066. https://doi.org/10.1029/2001JC001185

Song, X., Yu, L., 2017. Air-sea heat flux climatologies in the Mediterranean Sea: Surface energy balance and its consistency with ocean heat storage. Journal of Geophysical Research 122, 4068-4087. DOI:10.1002/2016JC012254

Sparnocchia, S., Schiano, M., Picco, P., Bozzano, R., Cappelletti, A., 2006. The anomalous warming of summer 2003 in the surface layer of the Central Ligurian Sea (Western Mediterranean). Annales Geophysicae 24, 443–452

Spitz, O., 2021. Rogue waves and extreme events, In: Mid-Infrared Quantum Cascade Lasers for Chaos Secure Communications, Springer Theses. Springer International Publishing, Cham, pp. 137–162. https://doi.org/10.1007/978-3-030-74307-9_5

Sprintall, J., Cronin, M.F., 2009. Upper ocean vertical structure, In: Encyclopedia of Ocean Sciences. Elsevier, pp. 217–224.

Stammer, D., Cazenave, A., Ponte, R.M., Tamisiea, M.E., 2013. Causes for contemporary regional sea level Changes. Annual Review of Marine Science 5, 21–46.

Stephens, G.L., Slingo, J.M., Rignot, E., Reager, J.T., Hakuba, M.Z., Durack, P.J., et al., 2020. Earth's water reservoirs in a changing climate. Proceedings of the Royal Society A: Mathematical, Physical and Engineering Sciences 476, 20190458.

Stewart, R.H., 2008. Introduction To Physical Oceanography. Texas A. and M. University, USA, 344pp.

Sudmeyer, R., Edward, A., Fazakerley, V., Simpkin, L., Foster, I., 2016. Climate change: Impacts and adaptation for agriculture in Western Australia (Report No. 4870). Department of Agriculture and Food, Western Australia, Perth, Australia, 177pp.

Sulisz, W., Paprota, M., Reda, A., 2016. Extreme waves in the southern Baltic Sea. Ciencias Marinas 42, 123–137.

Sultan, S.A.R., Elghribi, N.M., 2003. Sea level changes in the central part of the Red Sea. Indian Journal of Marine Sciences 32 (2), 114-122.

Sultan, S.A.R., Ahmad, F., El-Hassan, A., 1995. Seasonal variations of the sea level in the central part of the Red Sea. Estuarine, Coastal and Shelf Science 40, 1–8.

Sumner, G.N., 1977. Sea breeze occurrence in hilly terrain. Weather 32, 200–208.

Sutherland, G., Reverdin, G., Marié, L., Ward, B., 2014. Mixed and mixing layer depths in the ocean surface boundary layer under conditions of diurnal stratification. Geophysical Research Letters 41, 8469–8476. https://doi.org/10.1002/2014GL061939

Suthers, I.M., Young, J.W., Baird, M.E., Roughan, M., Everett, J.D., Brassington, G.B., et al., 2011. The strengthening East Australian Current, its eddies and biological effects-an introduction and overview. Deep Sea Research Part II: Topical Studies in Oceanography 58, 538-546.

Sutton, R.T., Hodson, D.L.R., 2005. Atlantic ocean forcing of North American and European summer climate. Science 309, 115-118.

Tadesse, M., Wahl, T., Cid, A., 2020. Data-driven modeling of global storm surges. Frontiers in Marine Sciences 7, 260. https://doi.org/10.3389/fmars.2020.00260

Talley, L.D., Pickard, G.L., Emery, W.J., Swift, J.H., 2011. Introduction to Descriptive Physical Oceanography. Elsevier, Boston, 550pp. https://doi.org/10.1016/B978-0-7506-4552-2.10001-0

Tebaldi, C., Strauss, B.H. and Zervas, C. E., 2012. Modelling sea level rise impacts on storm surges along US coasts. Environmental Research Letters 7, 014032.

Tekeli, A.E., Fouli, H., 2016. Evaluation of TRMM satellite-based precipitation indexes for flood forecasting over Riyadh City, Saudi Arabia. Journal of Hydrology 541, 471-479.

Thompson, P.R., Mitchum, G.T., Vonesch, C., Li, J., 2013. Variability of winter storminess in the Eastern United States during the Twentieth Century from tide gauges. Journal of Climmate 26, 9713–9726. DOI: 10.1175/JCLI-D-12-00561.1

Thomsen, M.S., Mondardini, L., Alestra, T., Gerrity, S., Tait, L., South, P., et al., 2019. Local extinction of bull kelp (*Durvillaea spp.*) due to a marine heatwave. Frontiers in Marine Science 6, 84. DOI: 10.3389/fmars.2019.00084

Thomson, J.A., Burkholder, D.A., Heithaus, M.R., Fourqurean, J.W., Fraser, M.W., Statton, J., et al., 2015. Extreme temperatures, foundation species, and abrupt ecosystem change: an example from an iconic seagrass ecosystem. Global Change Biology 21, 1463–1474.

Tintore, J., Gomis, D., Alonso, S., Wang, D.P., 1988. A theoretical study of large sea level oscillations in the Western Mediterranean. Journal of Geophysical Research 93, 10797–10803.

Tonbol, K.M., El-Geziry, T.M., Elbessa, M., 2018. Evaluation of changes and trends of air temperature within the southern Levantine Basin. Weather 73 (2), 60-66.

Trenberth, K.E., Jones, P.D., Ambenje, P., Bojariu, R., Easterling, D., Tank, A.K., et al., 2007. Observations: Surface and Atmospheric Climate Change, In: Climate Change 2007: The Physical Science Basis. Contribution of Working Group I to the Fourth Assessment Report of the Intergovernmental Panel on Climate Change. Cambridge University Press, pp. 235–336.

Triantafyllou, G., Yao, F., Petihakis, G., Tsiaras, K.P., Raitsos, D.E., Hoteit, I., 2014. Exploring the Red Sea seasonal ecosystem functioning using a three-dimensional biophysical model. Journal of Geophysical Research 119, 1791–1811. https://doi.org/10.1002/2013JC009641

Trujillo, A.P., Thurman, H.V., 2011. Essentials of Oceanography, 10th edition. Prentice Hall, Boston, 551pp.

Tsimplis, M.N., Álvarez-Fanjul, E., Gomis, D., Fenoglio-Marc, L., Pérez, B., 2005. Mediterranean Sea level trends: Atmospheric pressure and wind contribution. Geophysical Research Letters 32. https://doi.org/10.1029/2005GL023867

Ullmann, A., Moron, V., 2008. Weather regimes and sea surge variations over the Gulf of Lions (French Mediterranean coast) during the 20th century. International Journal of Climatology 28, 159–171. https://doi.org/10.1002/joc.1527

UNEP, 1996. State of the Marine and Coastal Environment in the Mediterranean Sea. Report No.100. United Nation Environmental Programme (UNEP), Athens, 142p.

Vilibic, I, Sepic, J., 2009. Destructive meteotsunamis along the eastern Adriatic coast: Overview. Physics and Chemistry of the Earth 34, 904–917. https ://doi.org/10.1016/j.pce.2009.08.004

Vilibic, I., Denamiel, C., Zemunik, P., Monserrat, S., 2021. The Mediterranean and Black Sea meteotsunamis: An overview. Natural Hazards 106, 1223-1267.

Vilibic, I., Sepic, J., Rabinovich, A.B., Monserrat, S., 2016. Modern approaches in meteotsunami research and early warning. Frontiersn in Marine Science 3, 57.

Vousdoukas, M.I., Voukouvalas, E., Annunziato, A., Giardino, A., Feyen, L., 2016. Projections of extreme storm surge levels along Europe. Climate Dynamics 47, 3171–3190.

Vucetic, T., Vilibic, I., Tinti, S., Maramai, A., 2009. The Great Adriatic flood of 21 June 1978 revisited: an overview of the reports. Physics and Chemistry of the Earth 34, 894–903.

Wahl, T., Chambers, D.P., 2016. Climate controls multidecadal variability in U.S extreme sea level records. Journal of Geophysical Research 121, 1274–1290. DOI:10.1002/2015JC011057

Wang, B., Ding, Q., Liu, J., 2011. Concept of global monsoon, In: The global monsoon system: Research and forecast. World Scientific, pp. 3–14. https://doi.org/10.1142/9789814343411_0001

Wang, C., 2018. A review of ENSO theories. National Science Review 5, 813-825.

Wang, C., Picaut, J., 2013. Understanding Enso Physics-A Review, In: Wang, C., Xie, S.P., Carton, J.A. (Eds.), Geophysical Monograph Series. American Geophysical Union, Washington, D. C., pp. 21–48. https://doi.org/10.1029/147GM02

Wang, C., Weisberg, R.H., Yang, H., 1999. Effects of the wind speed-evaporation-SST feedback on the El Nino–Southern Oscillation. Journal of the Atmospheric Sciences 56, 1391–1403.

Ward, B., Fristedt, T., Callaghan, A.H., Sutherland, G., Sanchez, X., Vialard, J., Doeschate, A., 2014. The Air–Sea Interaction Profiler (ASIP): An autonomous upwardly rising profiler for microstructure measurements in the upper ocean. Journal of Atmospheric and Oceanic Technology 31, 2246–2267. https://doi.org/10.1175/JTECH-D-14-00010.1

Warner, J.C., Armstrong, B., He, R., Zambon, J.B., 2010. Development of a Coupled Ocean–Atmosphere–Wave–Sediment Transport (COAWST) modeling system. Ocean Modelling 35, 230–244. https://doi.org/10.1016/j.ocemod.2010.07.010

Webb, P., 2021. Introduction to Oceanography. Roger Williams University, USA, 393pp.

Wells, N., 2011. The atmosphere and ocean: A physical introduction. 3rd Edition, Wiley, Hoboken, New Jersey, USA, 411pp.

Wertman, C.A., Yablonsky, R.M., Shen, Y., Merrill, J., Kincaid, C.R., Pockalny, R.A., 2014. Mesoscale convective system surface pressure anomalies responsible for meteotsunamis along the US East Coast on June 13th, 2013. Scientific Reports 4, 7143. https ://doi.org/10.1038/srep0 7143

Williams, A.P., Seager, R., Abatzoglou, J.T., Cook, B.I., Smerdon, J.E., Cook, E.R., 2015. Contribution of anthropogenic warming to California drought during 20122014. Geophysical Research Letters 42 (16), 6819-6828. https://doi.org/10.1002/2015GL064924.

Williams, D.A., Horsburgh, K.J., Schultz, D.M., Hughes, C.W., 2019. Examination of generation mechanisms for an English Channel meteotsunami: Combining observations and modeling. Journal of Physical Oceanography 49, 103-120. https ://doi.org/10.1175/JPO-D-18-0161.1

Wolf, J., Flather, R.A., 2005. Modelling waves and surges during the 1953 storm. Philosophical Transactions of the Royal Society A: Mathematical, Physical and Engineering Sciences 363, 1359-1375. https://doi.org/10.1098/rsta.2005.1572.

World Meteorological Organization (WMO), 2016. Guidelines on the Definition and Monitoring of Extreme Weather and Climate Events. Switzwerland, 62pp.

World Meteorological Organisation (WMO), 2017. Global Guide to Tropical Cyclone Forecasting (Report No. 1194). Switzerland, 397pp.

Ye, M., Li, S., Zou, Z., Song, J., He, H., Huang, J., Yang, H., 2020. Tower-based observation of air-sea momentum flux: comparisons between onshore and offshore winds. Acta Oceanologica Sinica 39, 61–68. https://doi.org/10.1007/s13131-020-1626-9

Yesubabu, V., Srinivas, C.V., Langodan, S., Hoteit, I., 2015. Predicting extreme rainfall events over Jeddah, Saudi Arabia: Impact of data assimilation with conventional and satellite observations. Quarterly Journal of the Royal Meteorological Society 142, 327-348.

Yim, S.-Y., Wang, B., Liu, J., Wu, Z., 2014. A comparison of regional monsoon variability using monsoon indices. Climate Dynamics 43, 1423–1437. https://doi.org/10.1007/s00382-013-1956-9

Yokoyama, S., Matsumoto, J., Araki, H. (Eds.), 2020. Nature, Culture, and Food in Monsoon Asia, International Perspectives in Geography. Springer Singapore, Singapore, 172pp. https://doi.org/10.1007/978-981-15-2113-3

Yoshikado, H., 1981. Statistical analyses of the sea breeze pattern in relation to general weather conditions. Journal of the Meteorological Society of Japan 59, 98–107.

Zavatarelli, M., Mellor, G.L., 1995. A numerical study of the Mediterranean Sea circulation. Journal of Physical Oceanography 25, 1384–1414.

Zebiak, S.E., 1993. Air–sea interaction in the Equatorial Atlantic region. Journal of Climate 6, 1567–1586.

Zervakis, V., Papadoniou, G., Tziavos, C., Lascaratos, A., 2003. Seasonal variability and geostrophic circulation in the eastern Mediterranean as revealed through a repeated XBT transect. Annales Geophysicae 21, 33–47. https://doi.org/10.5194/angeo-21-33-2003

Zhao, Y., Zhang, H., 2016. Impacts of SST Warming in tropical Indian Ocean on CMIP5 model-projected summer rainfall changes over Central Asia. Climate Dynamics 46, 3223–3238.

Zolina, O., Dufour, A., Gulev, S.K., Stenchikov, G., 2017. Regional hydrological cycle over the Red Sea in ERA-Interim. Journal of Hydrometeorology 18, 65-83.

https://www.bbc.com/news/world-middle-east-58783992.
"Shaheen: Tropical cyclone batters Oman and Iran, killing 13". BBC News on 04th October 2021. Retrieved 13th October 2021.